Enabling Metaverse and Telepresence Services in 6G Networks

RIVER PUBLISHERS SERIES IN COMMUNICATIONS AND NETWORKING

Series Editors:

ABBAS JAMALIPOUR
The University of Sydney, Australia

MARINA RUGGIERI
University of Rome Tor Vergata, Italy

The "River Publishers Series in Communications and Networking" is a series of comprehensive academic and professional books which focus on communication and network systems. Topics range from the theory and use of systems involving all terminals, computers, and information processors to wired and wireless networks and network layouts, protocols, architectures, and implementations. Also covered are developments stemming from new market demands in systems, products, and technologies such as personal communications services, multimedia systems, enterprise networks, and optical communications.

The series includes research monographs, edited volumes, handbooks and textbooks, providing professionals, researchers, educators, and advanced students in the field with an invaluable insight into the latest research and developments.

Topics included in this series include:

- Communication theory
- Multimedia systems
- Network architecture
- Optical communications
- Personal communication services
- Telecoms networks
- Wifi network protocols

For a list of other books in this series, visit www.riverpublishers.com

Enabling Metaverse and Telepresence Services in 6G Networks

Abdelhamied A. Ateya
Prince Sultan University, Saudi Arabia; Zagazig University, Egypt

Ahmed A. Abd El-Latif
Prince Sultan University, Saudi Arabia; Menoufia University, Egypt

Ammar Muthanna
SPbSUT, Russia

Artem Volkov
SPbSUT, Russia

Andrey Koucheryavy
SPbSUT, Russia

River Publishers

Routledge
Taylor & Francis Group
NEW YORK AND LONDON

Published 2025 by River Publishers
River Publishers
Alsbjergvej 10, 9260 Gistrup, Denmark
www.riverpublishers.com

Distributed exclusively by Routledge
605 Third Avenue, New York, NY 10017, USA
4 Park Square, Milton Park, Abingdon, Oxon OX14 4RN

Enabling Metaverse and Telepresence Services in 6G Networks / by Abdelhamied A. Ateya, Ahmed A. Abd El-Latif, Ammar Muthanna, Artem Volkov, Andrey Koucheryavy.

Routledge is an imprint of the Taylor & Francis Group, an informa business

ISBN 978-87-7004-673-2 (hardback)
ISBN 978-87-7004-785-2 (paperback)
ISBN 978-87-7004-675-6 (online)
ISBN 978-87-7004-674-9 (master ebook)

Contents

Preface

The modern world pulsates with the constant emergence of transformative technologies. Among them, the metaverse and telepresence services stand as beacons, promising to reshape the very fabric of human interaction. This book delves into the heart of these advancements, exploring their potential to revolutionize communication and unlock entirely new avenues for collaboration, learning, and social connection.

Telepresence offers a compelling solution for geographically dispersed individuals and communities. It transcends the limitations of distance, fostering a sense of real-time presence and interaction within virtual environments. Beyond mere video conferencing, telepresence strives to replicate the nuances of physical proximity, enabling richer and more immersive experiences. This powerful vision reaches its full potential when telepresence seamlessly integrates with the metaverse. This convergence has the power to usher in an era of unprecedented virtual connectivity, igniting a spark of creativity and innovation across all facets of society. By leveraging immersive technologies like augmented reality (AR), virtual reality (VR), and mixed reality (MR), the metaverse allows us to interact with digital worlds and each other in real time, fostering a sense of shared space and experience. However, the realization of this future hinges on advancements in communication technology. The emergence of 6G networks presents a pivotal turning point. By offering extreme bitrates, ultra-low latency, and exceptional reliability, 6G paves the way for the seamless transmission of vast amounts of data, a critical prerequisite for supporting sophisticated telepresence and metaverse experiences.

This book serves as a comprehensive guide, delving into the technical underpinnings of these technologies. Dedicated chapters explore the fascinating world of augmented and virtual realities, delve into the science of holographic images, and analyze the intricate workings of holographic technology. We investigate the role of key players like HTC Communications, and meticulously dissect the communication networks necessary for facilitating telepresence and holographic communication. The book further explores the development of test scenarios and methods for transmitting holographic images through network segments, alongside the research and development of subjective methods for assessing the quality of holographic image transmission. Finally, we delve into various methods for testing communication networks to ensure the seamless delivery of these immersive experiences.

Whether you are a seasoned technologist, a curious entrepreneur, or simply someone intrigued by the possibilities of the future, this book offers valuable insights and technical knowledge. It equips you with the tools to understand the intricate interplay between 6G networks, the metaverse, and telepresence services, and paves the way for your own contributions to this transformative journey. Let us embark on this exploration together and unveil the exciting future that awaits.

Abdelhamied A. Ateya,
Center of Excellence in Quantum and Intelligent Computing, Prince Sultan University, Riyadh, Saudi Arabia Zagazig University, Egypt

Ahmed A. Abd El-Latif,
Center of Excellence in Quantum and Intelligent Computing, Prince Sultan University, Riyadh, Saudi Arabia Menoufia University, Egypt

Ammar Muthanna,
Saint Petersburg State University of Telecommunications, Saint Petersburg, Russia

Artem Volkov,
Saint Petersburg State University of Telecommunications, Saint Petersburg, Russia

Andrey Koucheryavy
Saint Petersburg State University of Telecommunications, Saint Petersburg, Russia

Acknowledgement

The studies at St. Petersburg State University of Telecommunications by Prof. M.A. Bonch-Bruevich were supported by the Ministry of Science and High Education of the Russian Federation under the grant 075-15-2022-1137. This book also derived from a research grant funded by the Research, Development, and Innovation Authority (RDIA) – Kingdom of Saudi Arabia – with grant number (13325-psu-2023-PSNU-R-3-1-EF). Additionally, the authors would like to thank Prince Sultan University for their support.

List of Figures

List of Tables

List of Abbreviations

3GPP	3rd Generation Partnership Project
5G	Fifth-generation
6G	Sixth-generation
AI	Artificial intelligence
AR	Augmented reality
ASW	Asynchronous space warp
ATW	Asynchronous time warp
BER	Bit error rate
CDN	Content delivery network
COIN	Computing in the Network
CT	Computed tomography
D2D	Device-to-device
DetNet	Deterministic network
DS	Double stimulus
DSCQS	Continuous quality scale
DSIS	Distortion scale
DT	Test demonstration
eMBB	Enhanced mobile broadband
FC	Content fragment
FEC	Forward error correction
FoV	Field of view
GRP	Gross rating point
HDR	High dynamic range
HDTV	High-definition television
HMD	Head-mounted display
HolNetVerse	Holographic Network Verse
HTC	Holographic-type communication
IDE	Integrated development environment
IoT	Internet of Things
IP	Internet protocol
ITU	International Telecommunication Union

LAN	Local area network
LFV	Light-field video
LOD	Level of detail
M2M	Machine-to-machine
MAN	Metropolitan area network
ML	Machine learning
mMTC	Massive machine-type communication
MPEG-TS	MPEG Transport Stream
MR	Mixed reality
MRI	Magnetic resonance imaging
mULC	Massive ultra-reliable low-latency communication
NDI	Network Device Interface
NIR	Near-infrared
OBS	Open Broadcaster Software
P2M	Point-to-multipoint
P2P	Peer-to-peer
PBR	Physically based rendering
QoE	Quality of experience
QoS	Quality of service
QP	Quality parameter
QUIC	Quick UDP Internet Connections
RGB	Red, green, blue
RGB-D	Red green blue-depth
RIS	Reconfigurable intelligent surface
RTP-MPEGTS	Real-time Transport Protocol MPEG Transport Stream
RTT	Round trip time
SAMVIQ	Subjective assessment of multimedia video image quality
SC	Semantic communication
SC	Stimulus comparison
SD	Standard definition
SDN	Software-defined networking
SDSCE	Sources of simultaneous exposure
SDTV	Standard-definition television
SL	Structured light
SNMP	Simple network management protocol

SPbSUT	Saint Petersburg State University of Telecommunications
SS	Single stimulus
SSCQE	Single source continuous quality assessment
SUT	State University of Telecommunications
TCP	Transmission control protocol
ToF	Time of flight
TS	Test session
TSN	Time-sensitive networking
UDP	User datagram protocol
ULBC	Ultra-reliable low-latency broadband communication
uMBB	Ubiquitous mobile broadband
uRLLC	Ultra-reliable low latency communication
VLC	Visible light communication
WAN	Wide area network
WebRTC	Web real-time communication
WebRTC	Web Real-Time Communications
XR	Extended reality
XSS	Cross-site scripting

1

Introduction

1.1 Introduction

The modern world is constantly bombarded with new technologies that significantly impact every facet of human life and society. Among the most promising advancements shaping the future are the metaverse and telepresence services.

Telepresence offers a compelling solution for remote collaboration and communication. It empowers geographically dispersed teams, remote workers, and virtual communities to engage in interactive experiences [1]. These services have the potential to revolutionize communication network structures and user service delivery methods by enabling the creation of entirely new systems and applications [2].

At their core, telepresence and metaverse services leverage complex communication technologies to facilitate a sense of remote presence and interaction. Users gain the ability to participate in virtual experiences, events, and meetings from anywhere in the world [3]. Unlike traditional video or teleconferencing, telepresence strives to replicate the feeling of physical proximity, fostering more immersive and engaging remote interactions.

The future holds immense potential when telepresence seamlessly integrates with the metaverse. This convergence could lead to a surge in virtual connectivity, foster greater creativity, and drive innovation. These technologies offer groundbreaking opportunities to reimagine how we interact, learn, work, and socialize, ultimately paving the way for more inclusive, sustainable, and immersive digital societies. Within the metaverse, a shared virtual world, users can interact with digital environments and each other in real time using augmented reality (AR), virtual reality (VR), mixed reality (MR), and other immersive technologies [4].

The recent advancements in communication technology, coupled with the imminent arrival of next-generation cellular networks, are expected to significantly accelerate the implementation of telepresence and metaverse services. The rollout of fifth-generation (5G) technology ushered in a new era for telecommunication systems. As we move toward sixth-generation (6G), even stricter demands are anticipated to be met, with expectations of extreme bitrates, ultra-low latency, and ultra-high reliability for transmitted data [5]. This paves the way for supporting novel services like telepresence and the metaverse.

The emergence of 6G technology is poised to profoundly impact the future of telepresence and the metaverse. By delivering the necessary connectivity, performance, and intelligence, 6G networks have the potential to construct truly immersive, interactive, and interconnected virtual worlds, fundamentally transforming the landscape of telepresence and the metaverse [6].

1.2 Telepresence

Telepresence technology covers a collection of technological advancements designed to simulate physical presence or the sensation of being in a distant location. It facilitates real-time collaboration and interaction among individuals or organizations, disregarding their geographical separation [7]. By providing a more realistic and immersive experience, telepresence technology overcomes the limitations of conventional video conferencing.

The use of high-definition audio and video transmission is common, along with other features that enhance the impression of being remote. Telepresence systems use high-definition cameras to capture and transmit clear video. Multiple cameras are frequently utilized to present various views and ensure a complete picture of participants. Spatial audio technology is also used to generate the sensation of auditory presence. This entails recording and recreating audio with directional indications, allowing participants to determine the location and movement of sound sources [8]. Spatial audio enhances the realism of the telepresence experience, allowing for more natural and immersive talks.

Some telepresence systems incorporate robotic devices that can be controlled remotely. These devices, often equipped with cameras and sensors, allow users to navigate and interact with the remote environment. For example, a telepresence robot may be used to explore a distant location, attend meetings, or conduct inspections, giving users a physical presence in the remote setting [9]. A sense of contact is introduced into the telepresence experience via haptic feedback. It entails the utilization of tactile feedback devices capable of simulating pressure, vibration, or texture, among other stimuli. Haptic technology improves the functionality and authenticity of the telepresence environment through the provision of tangible feedback in response to the user's actions or interactions [10].

Telepresence solutions offer collaboration tools and capabilities that allow attendees to share papers, presentations, and multimedia content during virtual meetings. Screen sharing, whiteboarding, file sharing, and co-editing enable real-time collaboration and information exchange. Many fields can benefit from telepresence technology, including commerce, teaching, medicine, production, and the arts [11]. It opens a world of possibilities for telemedicine, virtual meetings, education, and remote collaboration – all of which rely on the sense of physical presence and real-time engagement.

Some telepresence systems use virtual avatars or representations of people to improve the sense of presence and identification during virtual meetings. Avatars can replicate their users' facial expressions, gestures, and body language, resulting in a more expressive and dynamic communication experience [12]. Telepresence services are frequently available on numerous devices and platforms, including desktop PCs, mobile devices, VR headsets, and web browsers. Cross-platform compatibility means that remote users can participate regardless of their chosen devices or operating systems.

1.3 Metaverse

The metaverse represents a new paradigm in human interaction, creativity, and expression, with endless opportunities for exploration, discovery, and creation. In the metaverse, virtual environments are persistent and continuously evolving, allowing users to explore, interact, and create within a shared digital space [13]. Virtual worlds encompass a wide variety of settings, including vast, empty environments and extremely detailed simulations of real locations.

The metaverse is based on interoperable platforms, allowing for seamless connectivity and communication across multiple virtual worlds and individuals. This interoperability promotes social engagement, cooperation, and content exchange across several virtual environments. Users play a central role in shaping the metaverse through content creation, customization, and personalization. From designing virtual avatars and building virtual objects to programming interactive experiences, users contribute to the richness and diversity of the metaverse ecosystem.

Immersive technologies such as virtual reality, augmented reality, haptic feedback, and spatial audio improve the metaverse's sense of presence and immersion, blurring the distinction between the physical and digital worlds

[14, 15]. Users can participate in realistic events, simulations, and adventures that go beyond standard media formats.

The metaverse relies heavily on social interaction, with users networking, conversing, and cooperating in shared virtual areas. Voice chat, text chat, avatars, and presence indicators are all social elements that help with real-time communication and interaction. Metaverse uses immersive technology to reimagine digital content consumption, collaboration, and interaction across a broad variety of sectors and use cases. This promising technology is expected to have a wide range of applications, including the following [16, 17].

- **Social networking and virtual communities:** Metaverse is a platform for social networking and virtual communities, allowing users to connect, communicate, and interact with one another in shared virtual environments. From virtual hangouts and social meetings to multiplayer games and virtual events, metaverse promotes meaningful social connections and community development across geographic boundaries.
- **Virtual meetings and collaborative workspaces:** Metaverse provides virtual meeting rooms and collaborative workspaces where users can hold meetings, presentations, and brainstorming sessions in immersive settings.
- **Entertainment and gaming:** Entertainment and gaming are key applications of the metaverse, with virtual worlds, games, and experiences that provide immersive storytelling, interactive gameplay, and shared adventures. Metaverse offers limitless entertainment and escapism alternatives, including virtual theme parks and immersive theater experiences, as well as multiplayer online games and virtual concerts.
- **Education and training:** Metaverse is used for educational reasons, including virtual classrooms, training simulations, and immersive learning experiences. VR and AR technologies provide interactive educational content, virtual field trips, and hands-on training simulations in a variety of disciplines, including science, history, healthcare, and professional skill development.
- **Virtual commerce:** Metaverse facilitates virtual commerce and retail experiences by allowing users to purchase, sell, and trade virtual goods, services, and assets within virtual environments. Virtual currencies, digital markets, and virtual storefronts promote economic transactions and consumer connections, opening new possibilities for e-commerce, digital art, and virtual fashion.

- **Digital marketing:** Brands and marketers use metaverse to run digital marketing campaigns, activate brands, and create immersive brand experiences. Virtual events, interactive commercials, and branded virtual spaces allow businesses to interact with their audiences in novel and memorable ways, increasing brand exposure, consumer engagement, and loyalty.
- **Healthcare and therapy:** Metaverse is used in healthcare applications such as telemedicine consultations, medical training simulations, and virtual therapy sessions. VR environments offer immersive experiences for patient education, rehabilitation, and mental health interventions, resulting in therapeutic benefits and improved patient outcomes.
- **Architectural design:** Architecture professionals use metaverse for virtual property tours, architectural visualization, and urban planning simulations. VR enables users to explore virtual properties, envision architectural concepts, and experience prospective developments before they are built, hence increasing decision-making and stakeholder participation in real estate projects.
- **Travel and tourism:** Metaverse provides virtual travel experiences and tourism destinations, allowing users to visit virtual reproductions of real-world monuments, attractions, and locations. VR tours, immersive storytelling, and interactive encounters enable people to enjoy virtual holidays and cultural experiences from the comfort of their own homes, encouraging tourism and destination marketing.

Telepresence and metaverse share many similarities; however, we can provide some differences between the promising technologies. Table 1.1 provides the main differences between telepresence and metaverse [7, 13, 18].

1.4 Challenges and Limitations of Metaverse and Telepresence

Telepresence technology provides numerous significant benefits in a wide range of fields. The integration of telepresence with metaverse technology opens in a new era of interaction and engagement, offering numerous significant benefits in various fields. In the field of business and commerce, these technologies revolutionize collaboration by exceeding physical boundaries, allowing seamless communication and cooperation among distant teams, partners, and clients. Telepresence and metaverse technologies provide instant interaction, overcoming geographical limitations. This leads to more

Table 1.1 Telepresence vs. metaverse.

	Telepresence	**Metaverse**
Definition	Telepresence refers to the use of advanced communication technologies to enable remote presence and interaction, allowing users to participate in virtual meetings, events, and experiences from anywhere in the world.	Metaverse refers to a collective virtual space that encompasses AR, VR, MR, and other immersive technologies. It represents a persistent, interconnected digital universe where users can interact with digital environments and each other in real time.
Scope	The objective of telepresence is to enable collaboration and communication from a distance by simulating the sensation of physical proximity and presence. Typically, it encompasses collaboration tools, audio communication, and real-time video conferencing to link users despite their physical separation.	Metaverse comprises an extensive variety of virtual environments, domains, and experiences. Within its expansive and interconnected virtual spaces, individuals can reside, labor, amuse themselves, and generate new content.
Applications	Telepresence is mainly employed in the context of virtual events, conferences, education, healthcare consultations, and conferences. This technology facilitates in-person collaboration and communication for individuals, remote teams, and remote laborers, eliminating the necessity for physical travel.	A wide range of applications, including social networking, amusement, gaming, education, commerce, and healthcare, make use of the metaverse. It provides opportunities for social interactions, collaborative endeavors, and immersive experiences across a wide range of industries and use cases.
Immersion and interaction	Telepresence facilitates collaboration and real-time communication by allowing users to engage in interactions with remote participants as if they were physically present in the same location. The objective is to accurately reproduce the subtleties of in-person communication, encompassing bodily movements, facial expressions, and spatial cognition.	Metaverse places significant emphasis on interactivity, wherein users actively participate in real-time communication, exploration, and creation within virtual environments. User-generated content creation, social interaction, and collaboration are fundamental components of the metaverse experience.

efficient decision-making, increased productivity, and the discovery of novel prospects for innovation and market growth. Moreover, they provide captivating virtual experiences including virtual showrooms, conferences, and product demonstrations, enhancing client involvement and fostering brand fidelity.

Beyond the business domain, telepresence and metaverse technologies are transforming social connection and community engagement. They provide platforms for virtual gatherings, social events, and online communities, facilitating interactions and relationships in a digital environment. People from various parts of the world can convene in virtual environments to cooperate, commemorate, and share ideas, surpassing geographical distances and cultural barriers. Furthermore, these technologies facilitate inclusivity and accessibility by offering remote participation alternatives, guaranteeing that anybody, irrespective of their geographical location or physical limitations, can engage in social activities, educational programs, and cultural events.

Telepresence and metaverse technologies have significant possibilities for service delivery and involvement in fields like healthcare, education, and entertainment. Virtual healthcare consultations, telemedicine services, and remote educational programs utilize these technologies to overcome geographical limitations, allowing individuals to access vital services and resources from any location worldwide. Similarly, immersive virtual experiences in gaming, entertainment, and cultural heritage preservation captivate audiences and enhance human experiences in extraordinary ways. Telepresence and metaverse technologies enable individuals and communities to prosper in a highly interconnected and digitally driven world by providing equal access to information, experiences, and opportunities.

Metaverse provides users with immersive and interactive experiences that go beyond conventional media forms, allowing them to explore, create, and engage in shared virtual spaces. It enables users to creatively express themselves through digital art, virtual environments, and interactive encounters, fostering innovation, self-expression, and artistic discovery. Metaverse offers global accessibility, allowing users to engage in virtual events, meetings, and activities without being limited by physical distance or geographical restrictions. Moreover, the metaverse facilitates social interaction, cooperation, and the establishment of communities, offering virtual environments for users to connect, communicate, and interact with others in real time.

Telepresence facilitates efficient collaboration among remote teams, irrespective of their physical distance, thereby promoting global teamwork and enhancing productivity. It has the potential to generate considerable financial benefits for organizations by eliminating the necessity for travel. These reductions may encompass expenditures associated with lodging,

Table 1.2 Limitations and challenges of telepresence [11, 19].

Challenge	**Limitation**
Telepresence systems require fast and dependable network connections to transmit high-quality audio and video in real time. Insufficient bandwidth or network congestion can cause decreased audiovisual quality, latency, or disruptions, affecting the entire telepresence experience. This reliance on network infrastructure might be a drawback, particularly in locations with poor or unreliable connectivity.	Telepresence systems are unable to perfectly recreate physical perceptions and experiences, restricting their ability to communicate tactile sensations, spatial awareness, and physical presence.
The user experience in telepresence environments may not entirely reproduce the complexity and depth of in-person interactions, making it difficult to transmit emotions, body language, and social indicators.	Telepresence is dependent on technology infrastructure and connectivity, making it vulnerable to disruptions, outages, and technical problems that can degrade user experience and reliability.
Telepresence systems may provide security threats, such as privacy breaches, data vulnerabilities, and unauthorized access to sensitive information, necessitating stringent security procedures and standards.	Telepresence environments may generate privacy concerns over data collecting, surveillance, and monitoring of user activities, generating inquiries about data ownership and consent.
Implementing telepresence systems may necessitate a large investment in infrastructure, including high-quality visual equipment, network infrastructure, and software platforms.	Integrating telepresence technologies with current workflows, systems, and processes may provide obstacles, necessitating organizational change management and training to guarantee smooth adoption and integration.

transportation, and organizational facilities. Moreover, telepresence facilitates the engagement of individuals who have limitations in mobility or disabilities in meetings, events, and activities, thereby fostering an environment that is more inclusive and accessible.

In addition to providing numerous advantages for collaboration, communication, and immersive experiences, telepresence and metaverse technologies face challenges and limitations concerning technology, user experience, privacy, and social impact. Table 1.2 provides the limitations and challenges of telepresence, while Table 1.3 provides the limitations and challenges of metaverse.

Table 1.3 Limitations and challenges of metaverse [16, 20].

<table>
<tr><th>Challenge</th><th>Limitation</th></tr>
<tr><td>Building and maintaining virtual environments in metaverse can be technically complex, requiring expertise in areas such as 3D modeling, game development, and virtual reality technologies.</td><td rowspan="2">Deploying metaverse systems may lead to significant costs, especially for top-tier solutions that include advanced cameras, audio devices, and display technology. In addition, the process of establishing and upkeeping the required equipment and infrastructure can be intricate and may necessitate specialist knowledge. The expense and intricacy can hinder mainstream acceptance, especially for small entities or individuals.</td></tr>
<tr><td>Metaverse communities need strong policies and tools for content moderation to handle problems like discrimination, and improper behavior, as well as to manage user-generated content.</td></tr>
<tr><td>Some users may face difficulties gaining access to the metaverse due to issues like low digital literacy, social inequities, and technological limitations. This could increase existing inequalities and lead to a digital gap among users.</td><td rowspan="2">Metaverse systems may not have the ability to accurately record complex non-verbal signals such as body language or facial expressions, which are essential for effective communication. Despite progress in this field, accurately replicating these signals in a remote environment might still pose difficulties. This constraint can affect the total communication encounter and might necessitate users to depend more on vocal indications.</td></tr>
<tr><td>Metaverse highlights issues of identification and privacy, such as digital identity, data ownership, and monitoring, generating discussions about user rights and protections in virtual environments.</td></tr>
</table>

1.5 Sixth-Generation Technology (6G)

6G communication networks hold significant promise for enabling telepresence and metaverse services by providing high-speed, low-latency connectivity, and supporting a wide range of use cases. From high-speed data transmission and low-latency communication to edge computing integration and AI-driven optimization, 6G networks have the potential to unlock new possibilities for immersive experiences across various domains [21].

The emergence of 6G communication technology signifies the subsequent progression in wireless connectivity, offering unparalleled levels of connectivity, speed, and dependability to satisfy the requirements of an ever more interconnected global society. Expanding upon the progress made in 5G, 6G endeavors to fundamentally transform communication networks through the implementation of innovative functionalities and capabilities that will reshape how individuals and technology engage. Figure 1.1 presents the main expected application and use cases of 6G technology as per the International

Telecommunication Union (ITU) and the 3rd Generation Partnership Project (3GPP) [22]. Telepresence and metaverse appear in the form of extended reality and holographic communications.

At its core, 6G is envisioned as a convergence of various emerging technologies, including artificial intelligence (AI), quantum computing, terahertz communication, and advanced materials. These technologies will enable 6G networks to operate at significantly higher frequencies, terabit-per-second data rates, and ultra-low latency, paving the way for transformative applications and services. One of the primary principles of 6G is its focus on ubiquitous connection to deliver continuous connectivity even in the most distant and difficult locations [23]. 6G networks will provide broad coverage and connection by utilizing improved antenna technologies, network densification, and satellite integration, enabling users to remain connected in any location. Figure 1.2 presents the main features of 6G networks, while Figure 1.3 presents the technical requirements.

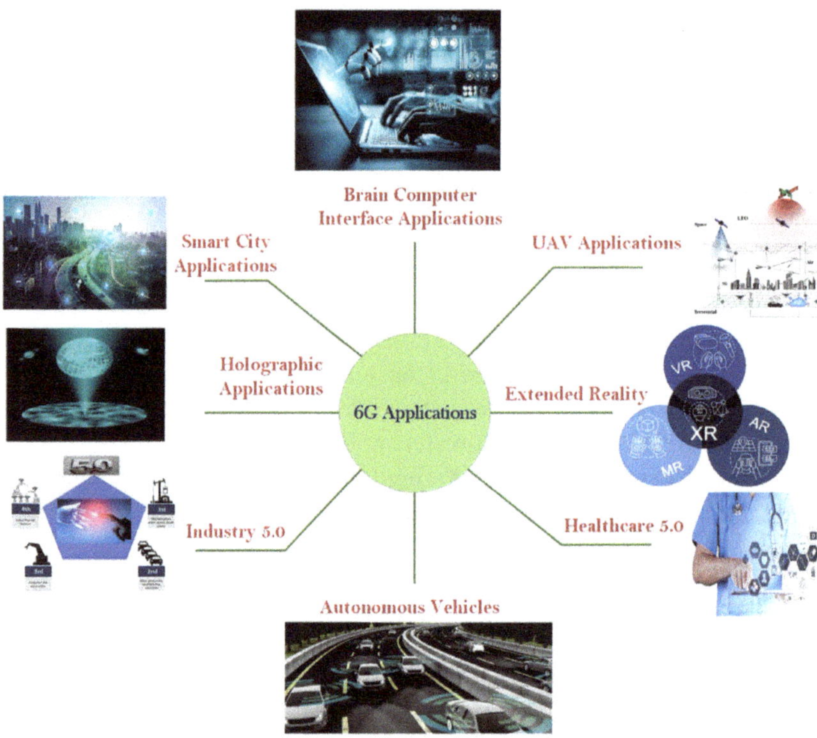

Figure 1.1 Main use cases of 6G technology.

Traditional communication networks, e.g., fourth-generation (4G) and fifth-generation (5G) cellular, have provided significant advancements in terms of connectivity and data speeds. However, the upcoming era of 6G networks presents new challenges and requirements that must be addressed to effectively support modern institutions' functions. These challenges include the need for ultra-low latency, massive device connectivity, high data rates, and enhanced security measures [25]. Table 1.4 summarizes the main features of 6G technology compared to 4G and 5G technologies [24, 26, 27].

The key aspects of 6G systems can be summarized as follows.

A. Connectivity and coverage

- **Extending coverage:** One of the primary benefits of 6G networks is the ability to extend coverage to areas where terrestrial infrastructure is limited or non-existent.
- **Ubiquitous connectivity:** 6G enables ubiquitous connectivity, meaning users can stay connected regardless of location. Users can access communication services seamlessly in urban areas,

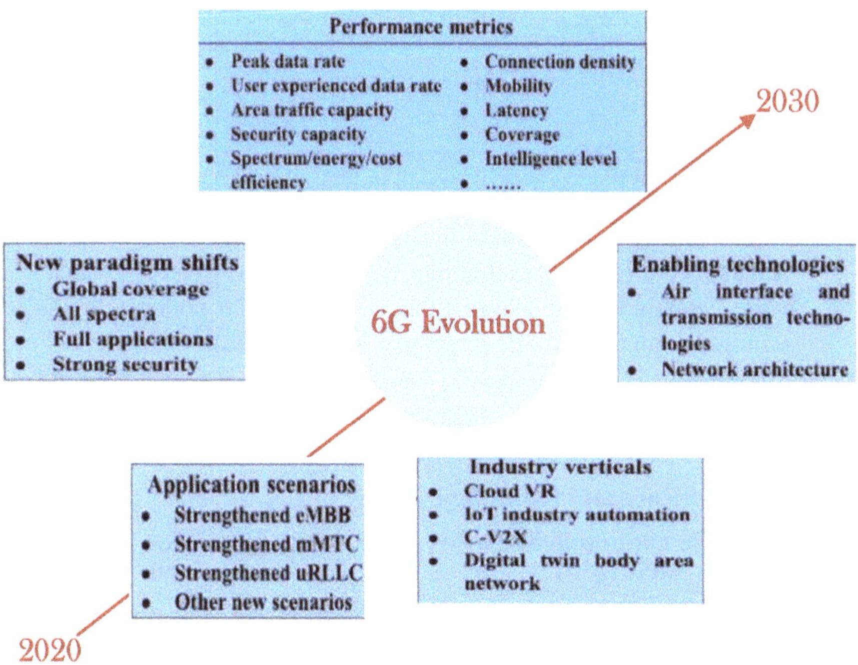

Figure 1.2 Main features of 6G technology [24].

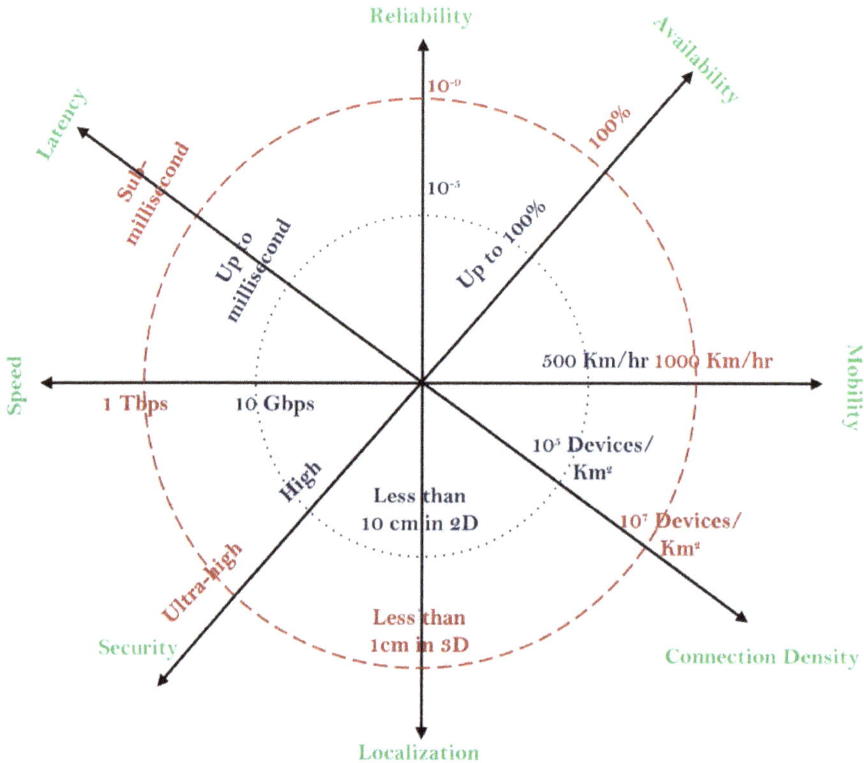

Figure 1.3 Main requirements of 6G technology [24].

suburban regions, or remote locations. This is particularly crucial for applications such as Internet of Things (IoT) devices, smart cities, and industrial automation, where reliable connectivity is required across diverse geographical locations [28].

- **Scalability:** As the number of connected devices and data traffic grows, integration ensures that the network can handle increasing demands, providing reliable connectivity to a larger user base.

B. Network resilience and redundancy

- **Reliable communication:** 6G systems should provide an ultra-high reliability of 10−9 [29].
- **Disaster recovery:** During natural disasters, such as hurricanes, earthquakes, or floods, terrestrial infrastructure may be damaged, making it challenging to maintain communication links. The 6G

Table 1.4 Main features of 4G, 5G, and 6G technologies.

Feature/ characteristic	4G Technology	5G Technology	6G Technology
Peak data rate	Up to 100 Mbit/s	Up to 10 Gbit/s	Expected > 100 Gbit/s
End-to-end latency	30–50 ms	1–10 ms	Sub-1 ms
Spectrum	Sub-6 GHz	Sub 6 GHz, 30–90 GHz, mmWave	Sub-THz, optical bands
Spectrum efficiency	Low (15 bps/Hz)	High (30 bps/Hz)	Ultra-high (100 bps/Hz)
Mobility	Medium (up to 350 km/hr)	High (up to 500 km/hr)	Ultra-high (up to 1000 km/hr)
Density of devices	Low-moderate	Moderate-high	High-ultra-high
Network slicing	Limited support	Extensive support	Advanced support
Edge computing	Limited support	Extensive support	Advanced support
MIMO	Up to 4 × 4 MIMO	Up to 64 × 64 MIMO	Expected to improve
Security and privacy	Secure, encryption	Enhanced security, network slicing isolation	Enhanced security, privacy, network slicing isolation, quantum-safe cryptography
Key technologies	LTE, OFDM	NR, massive MIMO, beamforming, network slicing	AI, THz communications, quantum networking
Use cases	Mobile broadband, IoT	Enhanced mobile broadband, IoT, critical communications	Immersive XR, holography, advanced automation

network provides different communication links that maintain communication even during a natural disaster [30].

- **Redundant connectivity:** Integrating satellite communications with 6G networks offers redundant connectivity options. Redundancy ensures that if one communication link fails or experiences disruptions, an alternative link can be utilized to maintain connectivity [31].
- **Remote and rural coverage:** One of the primary advantages of the 6G network is its ability to provide coverage in remote and rural areas where terrestrial infrastructure is limited or non-existent.

Therefore, the upcoming 6G networks are seeking to expand in a variety of ways, including longer battery life enabled by zero-energy gadgets, energy harvesting, and improved connectivity and coverage enabled by satellite and drone-coordinated communication. New technologies should be introduced to help the evolution of 6G integrated networks. AI is an intelligence-driven radio air network enabler. AI and machine learning (ML)-based approaches have emerged as a boon in a variety of domains, including image categorization, social networking, and security. AI/ML techniques will become increasingly important for RAN in terms of connectivity delivery, automation, and network-wide complexity management [32].

1.6 Book Vision

Since telepresence and metaverse are promising technologies that are expected to shape the future, this book provides the detailed technical aspects of both technologies. Telepresence services and metaverse can be integrated with virtual reality (VR) or augmented reality (AR) environments, allowing users to participate in virtual meetings and events set in immersive virtual worlds. VR telepresence allows users to interact with 3D surroundings and virtual objects, which increases engagement and cooperation. Also, holograms are the main base behind telepresence and metaverse.

Due to the potentials of hologram creation and holographic communications in telepresence and metaverse, we mainly consider the hologram creation and transmission in this book. Moreover, AR and VR technologies are considered as a part of telepresence and metaverse.

The book aims to provide professionals, researchers, and students in the field of telecommunications and networking with technical details on telepresence and metaverse. Mainly, the book provides technical briefs on technology creation and data transmission. Also, it provides directions for researchers working in the field of telecommunications and communication networks. It will be of interest to those seeking to understand the latest developments in 6G networks and their impact on metaverse and telepresence services.

References

[1] V. Souza, A. Maciel, L. Nedel, and R. Kopper, “Measuring presence in virtual environments: A survey,” *ACM Comput. Surv.*, vol. 54, no. 8, pp. 1–37, 2022.

[2] Y. Wang *et al.*, "A survey on metaverse: Fundamentals, security, and privacy," *IEEE Commun. Surv. Tutor.*, pp. 1–1, 2022.
[3] R. M. S. Jafar, W. Ahmad, and Y. Sun, "Unfolding the impacts of metaverse aspects on telepresence, product knowledge, and purchase intentions in the metaverse stores," *Technol. Soc.*, vol. 74, no. 102265, p. 102265, 2023.
[4] D. Wu, Z. Yang, P. Zhang, R. Wang, B. Yang, and X. Ma, "Virtual-reality inter-promotion technology for metaverse: A survey," *IEEE Internet of Things Journal*, 2023.
[5] K. A. Tychola, K. Voulgaridis, and T. Lagkas, "Tactile IoT and 5G & beyond schemes as key enabling technologies for the future metaverse," *Telecommun. Syst.*, vol. 84, no. 3, pp. 363–385, 2023.
[6] M. Adil, H. Song, M. K. Khan, A. Farouk, and Z. Jin, "5G/6G-enabled metaverse technologies: Taxonomy, applications, and open security challenges with future research directions," *J. Netw. Comput. Appl.*, vol. 223, no. 103828, p. 103828, 2024.
[7] D. M. Hilty *et al.*, "A review of telepresence, virtual reality, and augmented reality applied to clinical care," *J. Technol. Behav. Sci.*, vol. 5, no. 2, pp. 178–205, 2020.
[8] F. E. Fadzli, A. W. Ismail, and S. Abd Karim Ishigaki, "A systematic literature review: Real-time 3D reconstruction method for telepresence system," *PLoS One*, vol. 18, no. 11, p. e0287155, 2023.
[9] G. Zhang and J. P. Hansen, "Telepresence robots for people with special needs: A systematic review," *Int. J. Hum. Comput. Interact.*, vol. 38, no. 17, pp. 1651–1667, 2022.
[10] J. Feng, Z. Qian, G. Chen, and W. Wang, "A study on the importance of tactile stimulation on immersive experience for digital communication," in *Lecture Notes in Electrical Engineering*, Singapore: Springer Nature Singapore, 2022, pp. 393–401.
[11] R. Latifi, C. R. Doarn, and R. C. Merrell, *Telemedicine, telehealth and telepresence: Principles, strategies, applications, and new directions.* Cham, Switzerland: Springer Nature, 2020.
[12] D. Yang, J. Kang, T. Kim, and S.-H. Lee, "Visual guidance for user placement in avatar-mediated telepresence between dissimilar spaces," *IEEE Trans. Vis. Comput. Graph.*, vol. PP, no. 99, pp. 1–14, 2024.
[13] H. Wang *et al.*, "A survey on the metaverse: The state-of-the-art, technologies, applications, and challenges," *IEEE Internet Things J.*, vol. 10, no. 16, pp. 14671–14688, 2023.

[14] E. Dincelli and A. Yayla, "Immersive virtual reality in the age of the Metaverse: A hybrid-narrative review based on the technology affordance perspective," *J. Strat. Inf. Syst.*, vol. 31, no. 2, p. 101717, 2022.
[15] M. Xu *et al.*, "A full dive into realizing the edge-enabled metaverse: Visions, enabling technologies, and challenges," *IEEE Commun. Surv. Tutor.*, vol. 25, no. 1, pp. 656–700, Firstquarter 2023.
[16] S.-M. Park and Y.-G. Kim, "A Metaverse: Taxonomy, Components, Applications, and Open Challenges," *IEEE Access*, vol. 10, pp. 4209–4251, 2022.
[17] M. Ali, F. Naeem, G. Kaddoum, and E. Hossain, "Metaverse communications, networking, security, and applications: Research issues, state-of-the-art, and future directions," *IEEE Commun. Surv. Tutor.*, vol. PP, no. 99, pp. 1–1, 2023.
[18] S. Barta, R. Gurrea, and C. Flavián, "Telepresence in live-stream shopping: An experimental study comparing Instagram and the metaverse," *Electronic Markets*, vol. 33, no. 1, 2023.
[19] F. Edora Fadzli, A. Wanis Ismail, M. Yahya Fekri Aladin, and N. Zuraifah Syazrah Othman, "A review of Mixed Reality telepresence," *IOP Conf. Ser. Mater. Sci. Eng.*, vol. 864, no. 1, p. 012081, 2020.
[20] J. N. Njoku, C. I. Nwakanma, G. C. Amaizu, and D.-S. Kim, "Prospects and challenges of Metaverse application in data-driven intelligent transportation systems," *IET Intell. Transp. Syst.*, vol. 17, no. 1, pp. 1–21, 2023.
[21] C. De Alwis *et al.*, "Survey on 6G frontiers: Trends, applications, requirements, technologies and future research," *IEEE Open Journal of the Communications Society*, vol. 2, pp. 836–886, 2021.
[22] J. T. J. Penttinen, "On 6G Visions and Requirements," *J. ICT Stand.*, vol. 9, no. 3, pp. 311–325, 2021.
[23] H. Lee *et al.*, "Towards 6G hyper-connectivity: Vision, challenges, and key enabling technologies," *J. Commun. Netw.*, vol. 25, no. 3, pp. 344–354, 2023.
[24] C.-X. Wang *et al.*, "On the road to 6G: Visions, requirements, key technologies, and testbeds," *IEEE Commun. Surv. Tutor.*, vol. 25, no. 2, pp. 905–974, Secondquarter 2023.
[25] A. T. Jawad, R. Maaloul, and L. Chaari, "A comprehensive survey on 6G and beyond: Enabling technologies, opportunities of machine learning and challenges," *Comput. Netw.*, vol. 237, no. 110085, p. 110085, 2023.

[26] A. A. Ateya, A. Muthanna, M. Makolkina, and A. Koucheryavy, "Study of 5G services standardization: Specifications and requirements," in *2018 10th International Congress on Ultra Modern Telecommunications and Control Systems and Workshops (ICUMT)*, 2018, pp. 1–6.
[27] M. Alsabah *et al.*, "6G wireless communications networks: A comprehensive survey," *IEEE Access*, vol. 9, pp. 148191–148243, 2021.
[28] J. Wigard *et al.*, "Ubiquitous 6G service through non-terrestrial networks," *IEEE Wirel. Commun.*, vol. 30, no. 6, pp. 12–18, 2023.
[29] L.-H. Shen, K.-T. Feng, and L. Hanzo, "Five facets of 6G: Research challenges and opportunities," *ACM Comput. Surv.*, vol. 55, no. 11, pp. 1–39, 2023.
[30] A. Alhammadi, A. Abraham, A. Fakhreddine, Y. Tian, J. Du, and F. Bader, "Envisioning the future role of 3D wireless networks in preventing and managing disasters and emergency situations," *arXiv [cs.NI]*, 2024.
[31] D. Zhou, M. Sheng, J. Li, and Z. Han, "Aerospace integrated networks innovation for empowering 6G: A survey and future challenges," *IEEE Commun. Surv. Tutor.*, vol. 25, no. 2, pp. 975–1019, Secondquarter 2023.
[32] X. Lin, L. Kundu, C. Dick, and S. Velayutham, "Embracing AI in 5G-advanced toward 6G: A joint 3GPP and O-RAN perspective," *IEEE Commun. Stand. Mag.*, vol. 7, no. 4, pp. 76–83, 2023.

2

Augmented and Virtual Reality

2.1 The Concept of Augmented Reality and Its Application in the Modern World

Today, another type of service has appeared on the telecommunications market – augmented reality (AR) services. The main goal of augmented reality technology is not to create a new world but to improve the existing one by enhancing the user's senses such as hearing, vision, smell, etc. AR technologies are created based on International Telecommunication Union (ITU) recommendations, including J.301, 302, M.2083 [1]. According to ITU recommendation J.301, augmented reality refers to a type of mixed reality where graphical elements are integrated into the real world to improve the user experience and enrich information [2]. Mixed reality means the integration of virtual and augmented realities into human life.

Virtual reality completely replaces the existing world with an artificially formed environment, where all objects, without exception, are artificial, being inside this world. This means that a person does not feel or see in any way what is happening around him in the real world; he is completely immersed in another world. At the same time, augmented reality does not in any way create new worlds with new objects and relationships but only expands the existing reality. The main goal of augmented reality is to improve the user's interaction with the real world and provide him with additional information and capabilities. For example, when using AR devices to explore a new area, users can get additional information about attractions and local monuments, take pictures, post them on social networks, etc.

AR technologies are used in various fields and fields of activity, including the following [3]:

- **Advertising and marketing:** AR is used to create interactive advertising campaigns and allows potential consumers to look at a product from different angles.

- **Education:** AR is actively used to create interactive educational materials that allow you to visualize complex concepts and processes, which contributes to better learning by students. It allows the creation of interactive curricula and tests, educational applications, and programs for various disciplines and professions. Moreover, AR can help improve the efficiency of distance learning and meetings by allowing participants to connect in a virtual space and interact with each other more visually and effectively.
- **Tourism industry:** AR allows tourists to receive additional information about attractions and places of interest, guides to hotels, restaurants, and shops, as well as receive information about tickets and schedules using mobile devices.
- **Gaming:** AR technologies are used in various games, allowing players to interact with virtual objects in the real world. Industry and construction – AR is used to model and visualize various projects.
- **Automotive industry:** AR helps professionals design and build cars, and with AR applications, e.g., AR apps for driving, road users can receive additional information about their surroundings, which improves road safety.
- **Marketing:** AR helps provide consumers with a new and unique shopping experience by allowing them to try on virtual clothes, shoes, and accessories and see how the item fits with other items in their wardrobe before making a choice.
- **Entertainment industry:** AR can create unique and immersive content, enhance the usefulness of games, and satisfy user needs.
- **Medicine:** AR is used to create virtual training for medical students and to simulate surgeries for surgeons.
- **Industry:** AR has also found application in the industrial sector. It allows avoiding the risks of creating expensive physical prototypes of various devices and products, which significantly helps companies reduce their costs, launch new products in the market, and improve the quality of their existing products.

Overall, the use of augmented reality can significantly improve business and education opportunities in the modern world by bringing virtual experiences into the real world.

Augmented reality technology is still developing and improving, and the possibilities of its application in different areas are only expanding, which makes it promising for investors and businesses in general. Modern research confirms that AR is a popular technology and has great potential for application in various fields. There are many studies that show that AR is a relevant technology and can be useful in various fields. MarketsandMarkets conducted a study that expected the AR and VR industry to achieve revenue of $8.2 billion by 2028 [4]. Growth will be driven by increased demand for innovative technologies in various industries, such as healthcare, education, retail, and media.

There have been many developed proposals that show that AR can be an effective tool for improving worker productivity and saving resources in the industry. A study conducted by "PricewaterhouseCoopers International Limited (PwC)" found that VR and AR can help reduce design time by 60%, as well as reduce errors and improve workplace ergonomics. Another study in the medical industry was conducted by "Children's Hospital Los Angeles" and found that AR can help children better understand how the brain works, as well as help them cope better with medical procedures [5]. Therefore, AR remains a promising and in-demand technology for the present and the future. In the following sections, we will take a closer look at the use of AR in the modern world.

2.1.1 Augmented reality in medicine

The use of AR technology in medicine is widespread and has several advantages in solving specific problems for doctors. AR applications can be useful for helping people with disabilities, as these applications can be used for navigation to select the optimal route, considering the person's actual capabilities. AR applications are also being created to help restore hand motor skills in people who have lost coordination. The use of augmented reality in medicine is one of the most interesting and promising areas of development of this technology. It can be used in various procedures, ranging from training medical students to complex surgical procedures. Today, AR can be used in medicine in several key areas, including the following [6]:

1. Student education: AR is becoming increasingly popular in medical schools. With its help, students can study the anatomy of the human body in more detail, perform virtual surgeries, and learn various medical procedures. This allows them to gain additional practical experience, which is important for developing their professional skills.
2. Surgical support: AR can help doctors perform surgeries more accurately and safely for the patient. For example, "Engadget" reported on a project from "Novarad" that allows surgeons to see real-time

information about a patient's health and plan an operation. AR can also be used to visualize the surgical site and plan the procedure before it begins. The use of AR in surgery is not only an innovative issue but also a patient safety issue. Although there is still a lot to be done in this area.

3. Rehabilitation after injuries and operations: AR can help patients after undergoing treatment or surgery. For example, patients needing rehabilitation after an injury or surgery may be offered special applications with augmented reality. They can be used for various exercises and training that will help restore the function of the limbs and joints.
4. Diagnosis of diseases: AR can be used to diagnose diseases more accurately. For example, when conducting X-ray and ultrasound examinations, you can use special applications with augmented reality, which allow you to enlarge the image and see various anomalies and pathologies in more detail.

In general, the use of augmented reality in medicine is quite promising, as it helps improve the quality of procedures and operations and increases the accuracy of diagnosis and training of young specialists. The use of AR in surgery can significantly improve surgical outcomes, provide better visualization, and improve accuracy and safety.

AR has already proven to be a very reliable and useful medical tool. "Immersive Rehab" is a company that has developed rehabilitation programs using AR and VR for patients recovering from injuries and disabilities. "Immersive Rehab" programs have been designed for hospitals, rehabilitation centers, nursing homes, and home settings. They enable patients to recover faster by providing personalized and holistic rehabilitation that includes physical, cognitive, and psychological components. Programs also integrate modern technologies to help patients stop disease regression and promote functional recovery.

2.1.2 Augmented reality in the gaming industry

The authors of [7] investigated the impact of AR on the level of player engagement and cognitive performance. The study involved 149 participants who were divided into two groups. Each group played the same game, but one used AR, and the other only used 3D graphics. The study results showed that players in the group that used AR showed higher engagement with the game. They also showed higher levels of cognitive performance, as measured by the time taken to complete tasks and the number of errors made. Also, the authors get the following results:

- Players who used AR made 22% fewer errors than players in the group that only used 3D graphics.
- Task completion times were 2.3% slower in the group that used AR, but this difference was not statistically significant.
- Players from the AR group showed a higher tendency to repeat tasks and increased motivation to continue playing.

These results suggest that using AR in games can increase players' levels of engagement and improve their cognitive performance.

2.1.3 Augmented reality in tourism

The use of AR in tourism can enhance the tourist experience, improve understanding of cultural and historical heritage, help tourists better navigate a new place, and gain a more complete understanding of attractions and places. AR can be used to create virtual tours of museums, architectural monuments, parks, and other attractions [8]. Tourists can view augmented images over the actual location and learn more about the monuments, historical events, and other interesting places they encounter. AR can also help tourists find their location by providing additional directions and map directions on the phone. One of the benefits of AR in tourism is the ability to create personalized travel experiences. Tourists can choose what they want to see on the excursion, what format to choose, and how much time to devote to each attraction. They can also explore places that are not covered by traditional excursion programs. AR can be used in various tourism applications, including city walks, hiking trails, customized tours, museum exhibits, theme parks, and many others.

The "AccuWeather app" [9] is a popular weather app that uses AR to provide users with real-time information about climate conditions. One of the main features of the "AccuWeather app" is the AR feature, which allows tourists to know about the weather at their destination. It works as follows: the application displays the smartphone's camera and adds graphics with weather information virtually to the screen, which allows you to see the current weather in the place where the user is. To do this, you need to launch the AR camera and point it at the environment. The application will determine the user's location and show weather data for this location. For example, if the user is outside, the application will display the current temperature, wind speed, and chance of precipitation. However, the AR functionality of the "AccuWeather app" is not limited to just weather forecasts. It also provides information about air quality and the ozone index in the environment. The use of AR in the "AccuWeather" application in tourism helps tourists plan their trips and choose the most suitable time to visit a particular place.

There are many more applications using AR technology in the tourism sector. For example, "National Geographic Explorer" is an interactive application that allows tourists to immerse themselves in the virtual world of national parks and nature reserves [10]. The app uses AR technology to show many unique places that may not be known to most people. Users can select any location on the map and begin their virtual journey. They can open an AR window on their phone camera, through which they will see the world around them in a new way. The application displays information about the surrounding nature, animals, vegetation, and other objects. This app also allows tourists to get information about landmarks and other interesting places to visit in these national parks and sanctuaries. Users can select a route map and navigate from one point to another, learn about cultural and historical sites, explore flora and fauna, and visit other interesting places. "National Geographic Explorer" helps tourists feel as if they are in this beautiful and amazing environment, although, in fact, they are somewhere else. It also helps with education and allows you to share unique experiences with friends and family.

2.2 Communication Networks for AR Applications

When implementing network solutions, an important issue is to ensure high quality of service for each service, and the requirements for transmission parameters through the network will be different for different applications and types of traffic. ITU-T Recommendation Y.1540 identifies several of the most important network characteristics that affect the quality of service: throughput; reliability of the network/network elements; latency (ms) and latency jitter; amount of losses (%); network survivability – whether network elements can remain operational if individual elements fail [11].

For effective and uninterrupted use of augmented reality, it is necessary to provide a high-speed and stable communication network. This is necessary for fast data transfer and processing of information that is used to create animation and fill the virtual space. Powerful parallel computing and data transfer can require high network throughput, and have their own unique requirements for transmission quality, for example, minimal signal latency. Therefore, a communication network infrastructure is required that can meet these requirements and serve as a sufficiently stable and reliable data transmission channel.

Additionally, as the number of AR users increases, it becomes increasingly important to ensure that sufficient frequency channels are available and minimize external interference. This is necessary to avoid conflicts in network use and ensure maximum smoothness of data transfer. However, developing communication networks and increasing their efficiency may provide new opportunities for using augmented reality in business and other industries. Ultimately, the success of augmented reality in the market depends on how well-existing infrastructure and technologies can cope with network requirements, as well as how quickly new technological solutions will develop to help improve the quality of communication networks and eliminate their limitations.

Today, the existing network infrastructure, despite significant advances, has limitations in terms of throughput and communication stability. This is due to the fact that communication networks are needed to transmit large amounts of data, which are necessary for working with augmented reality.

In addition, the use of AR technologies requires most users to have high-speed internet access. Despite this, network technologies continue to evolve and improve to provide higher capacity and more resilient communications. For example, 5G wireless technologies have emerged that transmit data quickly and can improve the speed and quality of connections for AR users. In addition, communication network providers are constantly working to develop algorithms to optimize network traffic and provide more reliable communications to improve the quality of data transmission.

In general, the network infrastructure is constantly evolving, but for the successful use of augmented reality, it is necessary to further improve the technology and ensure sufficient bandwidth for data transmission for all users.

2.3 Virtual Reality

Virtual reality (VR) is an immersive technological innovation that empowers users to engage in interactive and realistic experiences within computer-generated environments. VR attempts to stimulate a sense of presence in users by accurately mirroring sensory experiences, including sight, sound, and contact, thereby transporting them to alternative worlds or scenarios. Head-mounted displays (HMDs), motion-tracking sensors, input devices (e.g., controllers), and software applications that generate and render virtual environments in real-time comprise the fundamental elements of VR.

In VR, users are immersed in a computer-generated simulation of a 3D virtual world and are able to move around and interact with it. By allowing the user to engage with the virtual world in real time and mimicking their physical presence, it aims to create a realistic and engaging sensory experience.

Virtual reality has the ability to enhance the user experience by incorporating several sensory inputs, not only visual ones. Some examples of this technology are haptic feedback systems that use vibrations or force feedback devices to mimic the sensation of touch, and spatial audio that realistically positions and moves sounds. These features make the virtual environment more engaging and participatory. VR technology is utilized in a wide range of fields, including architecture, healthcare, entertainment, education, training, and simulation. This technology empowers individuals to investigate digital environments and engage in interactions with virtual characters or objects. The main applications of VR are summarized as follows [12].

- **Gaming and entertainment:** One of the most widely used VR applications. VR gaming enables users to interact with objects and characters in real time while being completely immersed in virtual environments. In addition, entertainment venues utilize VR to create interactive experiences, virtual excursions, and immersive narratives.
- **Education and training:** VR can revolutionize the fields of education and training by providing immersive simulations that can be utilized to enhance architectural visualization, medical training, skill development, and historical reconstruction. It provides learners with the opportunity to participate in experiential learning and hands-on activities within a secure and regulated setting.
- **Healthcare:** VR is utilized in the healthcare industry for remote medical consultations, pain management, surgical training simulations, and patient rehabilitation. It serves as a powerful tool for both diagnosis and treatment, thereby improving healthcare delivery and patient outcomes.
- **Enterprise and industry:** VR is becoming more popular in areas such as architecture, engineering, manufacturing, and retail for design visualization, prototyping, product simulation, virtual prototyping, and customer interaction. It streamlines workflow processes, enhances collaboration, and lowers the expenses associated with physical prototypes.
- **Social interaction:** Social VR platforms allow users to engage with others in virtual worlds, promoting collaboration, communication, and social interaction in a virtual setting. These systems facilitate virtual gatherings, conferences, events, and shared experiences beyond geographic borders.

VR applications often include a variety of elements designed to provide users with immersive, engaging, and realistic experiences. Table 2.1 provides the typical features of VR applications [13, 14].

Table 2.1 Main features of VR applications.

Feature	**Discussion**
Heterogeneity	Capabilities of VR applications can vary depending on the platform, hardware, and software being used.
Realistic visuals/audio	VR applications aim for high-quality visuals and audio to improve immersion. The visual aspects strive to generate realistic and complex environments, textures, and objects, while the audio is intended to deliver spatial sound by accurately reproducing noises coming from various directions.
Immersive environment	VR provides a sense of immersion by surrounding the user with a 360° virtual environment. This immersive experience is often provided via HMDs, which cover the user's field of view and block off their real-world surroundings, allowing them to feel completely immersed in the virtual environment.
3D visualization/environment	VR environments are represented in 3D, allowing users to see depth and spatial relationships inside the virtual realm. This 3D representation adds reliability to the virtual environment and allows users to interact with items and situations from various perspectives. VR applications show users a computer-generated 3D environment that surrounds them, creating a sensation of presence and immersion. The environment may be static or dynamic, depending on the application.
Natural user interaction	VR applications are designed to give users intuitive and natural methods to engage with the virtual environment. This can involve hand gestures, vocal instructions, body movements, and even eye tracking. The idea is to give people a sense of presence and allow them to interact with the virtual world in a way that seems natural and immersive.
Multiplayer and social interaction	Many VR applications have multiplayer features that allow users to connect and collaborate with others in the same virtual world. This can include activities like multiplayer games, virtual meetings, and social networking in the virtual reality realm.
Real-time interaction	VR provides real-time interaction with virtual objects and surroundings, letting users to handle items, traverse around virtual locations, and participate in virtual experiences utilizing input devices such as motion controllers, hand tracking, or gesture.
Sensory feedback	VR technology attempts to mimic sensory sensations such as sight, hearing, and touch in order to generate a sense of presence in the virtual environment. This could include spatial audio to create realistic soundscapes, haptic feedback to replicate tactile sensations, and motion tracking to detect and respond to user movements.

2.3.1 VR applications for training and simulation

Virtual reality applications have revolutionized training and simulation across various industries by providing immersive, realistic, and interactive environments for learning and skill development. VR's immersive and interactive nature makes it an important tool in many industries for improving learning, skill development, and performance in secure and controlled conditions. VR-based training and simulation applications provide numerous benefits to the field of training, including improved trainee safety, increased engagement, and decreased training expenses. The potential of VR technology to revolutionize training and simulation in diverse sectors persists as it continues to advance. VR applications for training and simulation purposes include the following [15, 16]:

a. **Flight simulation:** Virtual reality is commonly utilized in aviation training to recreate realistic flight circumstances. Flight simulators that use VR technology provide pilots with realistic training environments in which pilots can practice flying various aircraft types, mastering flight controls, and dealing with emergencies. Pilots can practice takeoffs, landings, emergency procedures, and various flight techniques in a virtual environment that closely matches the cockpit and its surroundings. VR flight training is a cost-effective and safe alternative to traditional flight training, allowing pilots to get experience in a variety of weather situations and tough scenarios.
b. **Industrial and manufacturing training:** Virtual reality is used in the industrial and manufacturing sectors to train employees in equipment operation, safety procedures, and complex assembly processes. Users can practice tasks in a virtual environment that simulates real-world settings, reducing risk and increasing learning efficiency. VR is used in industrial training programs to help workers understand complex machinery, equipment functioning, and safety regulations in the manufacturing, construction, and heavy industries. VR simulations enable learners to practice tasks like equipment maintenance, assembly procedures, and emergency response drills in a virtual environment.
c. **Driving simulation:** Virtual reality driving simulators are used to train new drivers, evaluate driver performance, and undertake research on traffic safety and driver behavior. These simulators simulate real driving circumstances, traffic scenarios, and road surroundings to assist drivers improve skills including defensive driving, danger awareness, and vehicle control.

d. **Soft skills training:** Virtual reality is rapidly being utilized to develop soft skills such as communication, leadership, teamwork, and customer service. Immersive VR environments provide role-playing, scenario-based learning, and interpersonal interactions, allowing participants to practice and improve their soft skills through realistic simulations.
e. **Construction and architecture visualization:** VR is used in the construction and architecture industries to create virtual walkthroughs of building plans. This enables architects, engineers, and clients to view and interact with virtual representations of structures, assisting them in evaluating designs, identifying potential concerns, and making educated decisions.
f. **Military and defense training:** Virtual reality is widely utilized in military and defense training, allowing personnel to mimic battle scenarios, simulate tactical movements, and train in a variety of operating environments. VR can also be used to replicate vehicle and aircraft operations, weapons training, and virtual warfare scenarios. Furthermore, VR training systems enable soldiers to practice abilities such as shooting, battlefield navigation, and teamwork in realistic virtual situations without the risks associated with live exercises.
g. **Emergency response training:** Virtual reality is used to train emergency professionals, including firefighters, paramedics, and police officers. Simulations can include firefighting scenarios, disaster response training, medical emergency simulations, and crisis management exercises, allowing responders to practice key skills in a secure and realistic virtual environment.
h. **Sports training:** Virtual reality applications are utilized to monitor athlete performance, recreate game scenarios, and provide tailored coaching feedback. Athletes can use VR simulations to practice sports techniques, improve tactical decision-making, and visualize game strategy in a virtual sports arena or training facility.
i. **Medical training:** Virtual-reality-based medical simulators provide immersive training experiences for healthcare workers in surgical procedures, patient diagnosis, and medical interventions. Surgeons may practice complex surgeries in a virtual operating room, medical students can learn about anatomical structures in 3D, and emergency responders can replicate crisis circumstances to improve their skills and decision-making abilities.

Medical training is a common use case of VR applications in the field of training. VR is rapidly being used in medical education to simulate surgical

Table 2.2 VR applications in medical training.

Application	Discussion
Surgical procedure reproduction	VR enables the simulation of surgical procedures in a virtual environment. Detailed 3D models of anatomical parts, surgical instruments, and medical equipment are constructed to precisely depict the surgical setting. These models may be based on actual patient data, cadaveric anatomy, or generated anatomical models.
Immersive visualization of medical operations	By using a VR headset, trainees can immerse themselves in a virtual operating room and observe the surgical operation from any angle. They may examine the surgery site, instruments, and surgeon's operations in detail and realism. The immersive visualization enables learners to have a better knowledge of the technique and its complexities.
Medical scenarios variation and complexity	VR allows for the construction of several surgical situations with varied degrees of complexity. Trainees can proceed from basic operations to more difficult ones, progressively increasing their abilities and confidence. Furthermore, VR can imitate problems and unforeseen occurrences that may occur during surgery, allowing trainees to practice problem-solving and decision-making skills.
Performance assessment and evaluation	VR systems can collect information about trainees' performance during surgical simulations. Metrics such as time taken, movement precision, and procedural adherence can all be recorded and examined. This information can be utilized to provide objective assessment, feedback, and evaluation of trainee progress, highlighting areas for development.
Haptic feedback	Some advanced VR systems feature haptic feedback devices that give users a sense of touch and force. Trainees can use specialized haptic controllers or gloves to experience the sensation of grasping surgical instruments and interacting with tissues. This adds a haptic component to the simulation, increasing the realism of the experience.
Collaborative learning	Virtual reality can facilitate collaborative learning by allowing numerous trainees to engage in the same surgical simulation. Trainees can interact and collaborate electronically, emulating a surgical team setting. This collaborative method encourages cooperation, communication, and coordination skills, all of which are important in real-world surgical settings.
Skill development and practice	Virtual reality allows learners to practice surgical skills repeatedly in a safe and controlled setting. They can do virtual incisions, suturing, tissue manipulation, and other surgical procedures. The system can provide real-time feedback on the trainee's performance, including parameters like precision, speed, and accuracy, enabling ongoing improvement and skill development.

operations and create a realistic learning environment for medical students and professionals. Table 2.2 provides the main aspects of deploying VR in medical training and how the promising technology shapes the future of the medical field. VR simulations can be easily customized to simulate surgical procedures [17, 18]. The adaptability of VR technology enables the development of personalized simulations that closely mimic real-world surgical experiences. Customizing VR simulations allows educators and surgeons to construct extremely specialized and targeted training experiences that closely correlate with the surgical procedures and scenarios they want to reproduce. This modification increases the instructional value of the simulations and offers learners a more realistic and effective learning environment. Table 2.3 provides the main directions of VR in simulating medical surgeries [19, 20].

Table 2.3 VR applications in medical surgeries.

Application	Discussion
Surgical pathways and techniques	VR simulations have the capability to mimic particular surgical pathways and techniques that are integral to a given procedure. For each phase of surgery, educators and surgeons can define and program the necessary instrument interactions, movements, and steps. This customization guarantees that trainees acquire knowledge of the precise techniques employed in the targeted surgery and engage in meticulous practice of the procedural steps.
Complication scenarios	VR simulations can integrate a multitude of complications and unforeseen occurrences that might transpire throughout the surgical procedure. It is possible to program challenges such as bleeding, tissue damage, anatomical variations, or equipment malfunctions into these scenarios. Trainees can cultivate critical thinking abilities, hone their capacity to respond appropriately to unforeseen circumstances, and enhance their adaptability through the introduction of complications.
Patient-specific anatomy	By utilizing medical imaging data such as magnetic resonance imaging (MRI) or computed tomography (CT) scans, VR simulations can incorporate patient-specific anatomy. This enables the development of virtual models that closely illustrate the anatomical features of the patient, encompassing pertinent organs, tissues, and structures that are necessary for the surgical intervention.
Instrumentation and equipment	A variety of surgical instruments and equipment that are particular to the procedure being simulated may be incorporated into VR simulations. Surgical instruments such as scalpels, forceps, retractors, suturing tools, and others are encompassed within this category. By precisely simulating the behavior of the virtual instruments, trainees can gain practical experience with the precise tools necessary for the procedure.

Table 2.3 *Continued.*

Application	Discussion
Feedback and evaluation	VR simulations can provide trainees with immediate feedback and evaluation. Metrics such as accuracy, precision, speed, and adherence to procedural steps may be incorporated into this feedback. Immediate feedback on performance enables trainees to pinpoint areas that require enhancement and hone their abilities.
Realism and accuracy	Creating a high level of realism and accuracy in VR simulations can be difficult. While efforts are made to generate accurate anatomical models and reproduce surgical processes, there may be limits in portraying the nuances and complexities of actual surgical circumstances. Ensuring that the simulation accurately replicates the real-world experience necessitates meticulous attention to detail and ongoing refinement.
Real-time physiological modeling	Use real-time physiological modeling in VR simulations to represent anatomical structures' behavior and responses, such as tissue deformation, hemorrhage, or fluid dynamics. Realistic physics-based simulations can help to better reflect the interactions between surgical equipment, tissues, and fluids.

2.4 VR Rendering

VR rendering is the process of developing and showing visual content in VR settings, resulting in immersive and realistic virtual experiences for users. VR rendering requires numerous critical components and approaches to enable seamless, high-quality images, and optimal performance. Virtual reality rendering includes the following five main processes.

1. **Scene geometry and modeling:** The initial stage in VR rendering is to create or import 3D models and scene geometry, which constitute the virtual environment. Scene geometry refers to the virtual world's objects, surfaces, textures, lighting, and other features [21].
2. **Texturing:** Texturing is the process of applying textures and materials to 3D objects to enhance detail, realism, and visual richness in the virtual environment. Textures can imitate surface attributes like color, reflectivity, roughness, and bumpiness by using diffuse maps, specular maps, normal maps, and other texture types. Material qualities like reflectance, transparency, and emissiveness are established to accurately render the appearance of virtual objects in various lighting circumstances [22].
3. **Lighting and shading:** Lighting is vital in VR rendering because it illuminates the virtual world and creates realistic shadows, highlights,

and reflections. Several lighting approaches, including global illumination, ambient occlusion, and dynamic lighting, are used to replicate natural lighting effects and improve visual realism. Surface shading and specular highlights are correctly calculated using shading models such as Phong shading, Blinn-Phong shading, and physically based rendering (PBR) [23, 24].

4. **Render pipeline:** The rendering pipeline is divided into various stages: geometry processing, rasterization, shading, and post-processing. Geometry processing entails converting 3D objects to screen-space coordinates and executing culling and clipping procedures to improve rendering performance [25]. Rasterization turns geometric primitives (such as triangles) into pixels on the screen, determining which pixels are visible and which are obscured. Shading uses vertex and fragment shaders to determine the color and intensity of each pixel based on lighting circumstances, surface attributes, and material parameters. Anti-aliasing, depth of field, motion blur, and tone mapping are examples of post-processing effects used to improve image quality and visual appeal.
5. **Stereo rendering and optimization:** In VR rendering, stereo rendering algorithms are utilized to generate separate images for each eye, creating a sense of depth and stereoscopic vision. Optimization techniques like occlusion culling, level of detail (LOD) management, and GPU instancing are used to increase rendering speed and keep frame rates steady for a smooth VR experience. Asynchronous time warp (ATW) and asynchronous space warp (ASW) techniques are used to reduce latency and motion sickness by anticipating and compensating for head movement in real time [26].

Several advanced rendering techniques are used to improve visual realism in VR simulations. This includes the following [27–29]:

a. **Physically based rendering (PBR):**PBR is a rendering technology that accurately models light's interaction with materials. It considers surface qualities, reflectivity, roughness, and subsurface scattering to produce more realistic and visually appealing virtual worlds. PBR may substantially improve the appearance of materials, making them more realistic and lifelike.
b. **Global illumination:** Global illumination approaches mimic indirect lighting in a scene, including diffuse inter-reflection, color bleeding, and soft shadows. Global lighting improves the realism of VR simulations

by precisely modeling how light bounces off objects and illuminates the scene, as well as perception of depth and spatial relationships.

c. **Screen-space reflections:** Screen-space reflections replicate the reflection of objects based on their visibility on the screen. This method approximates reflections by sampling the available screen space, resulting in more efficient and real-time depiction of reflecting surfaces. It improves the visual quality of reflective elements like mirrors and polished surfaces while also increasing the overall realism of the virtual world.
d. **High dynamic range (HDR) rendering:** Unlike traditional rendering approaches, HDR renders a larger range of brightness values, resulting in more realistic and visually appealing scenes. It accurately represents the contrast between bright and dark areas, resulting in greater lighting realism and detail preservation throughout the virtual world.
e. **Anti-aliasing:** Anti-aliasing techniques are used to eliminate jagged edges and smooth the display of curved or diagonal lines. Anti-aliasing increases the visual quality and realism of VR simulations by reducing pixel distortion, resulting in smoother and more natural-looking simulated pictures.
f. **Subsurface scattering:** Subsurface scattering is a rendering approach that simulates how light enters and scatters within translucent objects like skin, flesh, and leaves. Subsurface scattering improves the appearance of soft tissues and organic objects by precisely modeling how light interacts with these materials, making them more alive and visually convincing.
g. **Volumetric rendering:** Volumetric rendering is a technique for representing and drawing three-dimensional data volumes such as smoke, fog, or interior organs. Volumetric rendering enhances the virtual environment's depth and realism by simulating light interactions with these volumes. It is very useful for rendering inside structures and tissues in medical VR simulations.

These advanced rendering techniques, along with others, help to improve the visual realism and authenticity of VR simulations. By adding these techniques into the rendering pipeline, virtual worlds become more immersive, visually appealing, and closely reflect real-world scenarios, improving the entire training experience for surgical simulations.

Real-time rendering engines, such as Unity and Unreal Engine, have made a substantial contribution to the widespread use of PBR in VR simulations. These engines offer user-friendly interfaces, sophisticated rendering capabilities, and substantial development resources, allowing developers and artists to build visually accurate and immersive VR experiences that are closely related to real-world surgical scenarios. Real-time rendering engines used in VR simulations offer a range of important features. Table 2.4 provides the main features of real-time rendering engines.

Table 2.4 Main features of real-time rendering engines.

Feature	**Issue with real-time rendering**
Stereoscopic rendering	Real-time rendering engines use stereoscopic rendering techniques to produce separate views for each eye, resulting in the illusion of depth and 3D immersion. They manage the required transformations and rendering optimizations to provide proper stereoscopic rendering, which improves the sense of presence in VR simulations.
Spatial audio	Real-time rendering engines offer spatial audio approaches, which allow for an immersive audio experience in VR simulations. They offer tools and APIs for spatializing audio sources based on the user's location and orientation, emulating realistic sound propagation, directionality, and distance attenuation. Spatial audio improves total immersion and presence in the virtual environment.
Animation	Real-time rendering engines include powerful animation and character systems that bring virtual characters and objects to life. They offer tools for skeletal animation, inverse kinematics, facial animation, and blending forms. These systems enable realistic character movements, expressions, and interactions, which improves the credibility and immersion of VR simulations.
Networking and multiplayer support	Real-time rendering engines enable numerous users to communicate and contribute in the same virtual world. They manage network synchronization, authoritative server architecture, and latency compensation to enable multiplayer VR experiences, making VR simulations appropriate for team training or collaborative scenarios.
Content creation and workflow tools	Real-time rendering engines provide content production tools and procedures that simplify the process of building VR simulations. They offer visual editors, asset importers, scripting languages, and debugging tools to help developers and testers create and iterate VR experiences. These technologies allow developers and artists to create and iterate on VR content more efficiently.
Physical simulation	Real-time rendering systems frequently contain physical simulation features for objects and interactions in the VR environment. They support realistic object dynamics, collision detection, and physics-based interactions, enabling for more realistic interactions between virtual items, actors, and the environment.

Table 2.4 *Continued.*

Feature	Issue with real-time rendering
Performance optimization	Real-time rendering engines focus performance optimization to provide fluid and responsive VR experiences. They use a variety of approaches, including occlusion culling, level-of-detail systems, and efficient rendering pipelines, to increase frame rates and reduce latency. These enhancements are crucial to ensuring a pleasant and engaging VR experience.
VR input and interaction	Real-time rendering engines can handle VR input devices including motion controllers, haptic feedback devices, and eye-tracking systems. They provide APIs and tools for mapping human input to in-game activities, enabling interactive experiences in which users can manipulate items, execute gestures, and explore the virtual environment.

2.5 Communication Networks for Virtual Reality Applications

Communication networks play an important role in enabling VR applications by simplifying data transmission between VR devices, servers, and content suppliers. Effective communication networks are essential for providing immersive and seamless VR experiences to users in a variety of contexts and use cases, as they ensure reliable connectivity, low latency, and high-quality content delivery. As VR technology evolves, developments in communication networks will be critical in improving the performance and accessibility of VR applications.

To achieve the required experiences of VR applications, communication networks for such applications should deploy novel network structures and technologies. Table 2.5 provides the main challenges that face the development of communication networks for VR applications [30, 31]. Addressing such communication network challenges is critical for providing immersive and smooth VR experiences to people worldwide. Continued research and development in networking technologies, combined with industry collaboration and standardization efforts, will drive the advancement and adoption of virtual reality in the future.

Heterogeneous communication networks and interfaces can be used to support VR applications. This includes existing network technologies and the upcoming networks, including 6G. Table 2.6 summarizes the communication technologies that can assist VR applications [31, 32].

Table 2.5 Challenges of communication networks for VR applications.

Challenge	Discussion
Bandwidth	VR applications rely on high-bandwidth, low-latency communication networks to ensure real-time rendering and interactivity.
Latency	Delay in data transfer can cause motion sickness, disorientation, and diminished immersion. Achieving continuous network performance is critical for providing a smooth VR experience. VR applications require an end-to-end latency of 5 msec.
Content distribution	The distribution of large-scale VR content, such as high-resolution 3D models and immersive films, presents storage, bandwidth, and scalability difficulties. Content delivery networks (CDNs), edge computing, and peer-to-peer (P2P) networks are used to improve content distribution and minimize latency in VR applications.
Interoperability and standardization	Ensuring interoperability and compatibility between VR hardware devices, software platforms, and communication protocols is critical for seamless integration and widespread acceptance of VR applications. Standardization efforts, corporate collaborations, and open-source initiatives all seek to build common frameworks and protocols for VR development and deployment.
Quality of service (QoS)	VR applications require dependable and consistent QoS measurements such as throughput, packet loss, and jitter. Variability in network performance can affect the quality of VR experiences, resulting in visual distortions, audio distortion, and user discomfort. Network optimization and prioritizing algorithms are required to meet the QoS requirements for VR applications.
Network congestion	High network congestion can reduce VR performance and increase delay, particularly in crowded areas or during high usage hours. Traffic management techniques, such as traffic shaping, congestion control, and dynamic bitrate adaptation, are used to reduce congestion and ensure a steady network connection for VR users.

Table 2.6 Communication technologies for VR applications.

Network	Aspects
Local area networks (LANs)	• LANs are often utilized in VR applications deployed within a specific area, such as homes, companies, or universities. • LANs enable high-speed, low-latency communication, allowing VR material to be seamlessly transmitted across devices on the same network. • LANs are ideal for multiplayer VR experiences, collaborative workspaces, and shared multimedia streaming in constrained physical environments.

Table 2.6 *Continued.*

Wi-Fi	• Wi-Fi networks provide wireless connectivity for VR devices, allowing users to access web content, download updates, and stream multimedia content. • Wi-Fi networks give VR users freedom and mobility, allowing them to roam around without being linked to a physical connection. • Advanced Wi-Fi protocols, such as Wi-Fi 6 (802.11ax), provide better speed, lower latency, and more dependability, making them ideal for demanding VR applications.
Peer-to-peer (P2P)	• P2P networks enable decentralized content distribution and collaborative experiences by facilitating direct communication and data sharing between VR devices. • P2P networks can improve the scalability and defect tolerance of virtual reality applications while decreasing reliance on centralized servers. • VR applications frequently employ P2P networking protocols, including WebRTC (Web real-time communication), to facilitate user-to-user content sharing, multiplayer gaming, and real-time communication.
Internet of Things (IoT)	• VR applications may employ IoT networks to facilitate communication with intelligent objects or sensors situated in the physical environment. • IoT networks may be integrated into VR training simulations, enabling users to engage with physical medical devices or equipment. • IoT networks facilitate the integration of data and interactions from the physical world into the virtual environment.
Satellite networks	• Satellite networks offer connectivity for VR applications in geographically isolated or rural regions devoid of terrestrial networks. • Users in underserved areas or while engaged in outdoor activities can access immersive content, online entertainment, and virtual experiences through satellite-based VR services.
5G/6G networks	• Cellular networks offer VR devices wireless connectivity in regions with limited or nonexistent Wi-Fi coverage. • By providing extensive coverage, low-latency communication, and rapid data transmission, 5G and 6G cellular networks enable virtual reality applications such as remote collaboration, live streaming, and multiplayer gaming to function.

Table 2.6 *Continued.*

	• 5G/6G networks have the potential to significantly improve the delivery of immersive VR experiences by enabling massive machine-type communication (mMTC) and ultra-reliable low-latency communication (uRLLC). This advancement would enable high-quality content delivery and real-time interaction.

2.6 Practical Use Cases of AR and VR

A very good example of the use of AR is a study published in the *Journal of Educational Computing Research* [33]. It was held in the People's Republic of China in 2018. The study aimed to explore how the use of AR in the learning process can increase student motivation and improve learning outcomes. The researchers developed a mobile app that used AR to help students identify the parameters of stars in the night sky. The application has a built-in star search function and a function to open additional information about the stars, such as location, characteristics, name, etc. The study was conducted on a group of students aged 16–18 years. The control group studied using traditional methods, and the experimental group used a mobile application with AR.

The study results showed that using AR in the educational process significantly increases student motivation and improves learning outcomes. Students in the experimental group showed more interest in the subject, asked more questions, and showed a higher level of understanding of the topic. Thus, the study confirmed that the use of AR in education can positively affect learning by increasing student motivation and improving their learning outcomes.

The medical company "Novarad" has developed a virtual reality system [34] in which the surgeon can view the patient in real time and receive information about the patient's health. It can also combine a patient's image with medical data such as X-rays and scans into a digital mask. This helps doctors more accurately determine the location of the operation and understand which tissues they can focus on.

Immersive Rehab programs use motion sensors and virtual reality controllers to allow patients to control their movements and monitor their progress. During rehabilitation, patients work in a playful environment, which promotes better mood, engagement, and motivation. The company's main goal is to help patients recover quickly, improve the effectiveness of rehabilitation, and reduce problems such as pain and movement restrictions. Here are some examples of rehabilitation programs that Immersive Rehab creates.

- **River Ranger:** A rehabilitation program for stroke and brain injury survivors. It allows patients to improve balance, coordination, and strength in the lower extremities.
- **Time Walk:** A rehabilitation program for patients with neuromuscular diseases such as Parkinson's disease or cerebral palsy. Patients improve their reach zone, coordination, and balance.
- **Skate Away:** A rehabilitation program for patients after injury or surgery to the legs or lower legs. Patients train their muscles, increase flexibility, and get rid of pain.

Despite the many benefits of using AR and VR, we should also pay attention to the limitations associated with its use. First, the cost of hardware and software to use AR/VR can be high, limiting the technology's accessibility to a wider audience. Second, some people may experience discomfort when using AR/VR due to sensory overload or balance issues. Additionally, technology use can cause eye strain and headaches. Also, the use of AR/VR may entail certain legal and ethical issues related to the protection of users' personal data and property rights.

2.7 Summary

Augmented reality is a technology that allows adding virtual elements to the real-world using computer vision and various sensors. Today, augmented reality is widely used in various fields, including education, medicine, advertising, entertainment, architecture, and many others. Its use can transform our world and significantly contribute to modern culture and economy. AR has already proven its effectiveness in training and education, allowing students to gain unique and visual experiences and improving the efficiency of equipment maintenance in various industries. In medicine, augmented reality can be used both by doctors to diagnose and treat patients and with patients who, thanks to this technology, have learned to manage their health better. In education, augmented reality can be used to improve the quality of learning and increase student motivation. Thus, students can gain unique experience and knowledge using augmented reality in the educational process. In manufacturing and engineering, augmented reality can help visualize and distribute production processes, reduce risks, and optimize business processes.

Commercially, AR has become a popular means of promoting products and services, providing highly engaging and brand-evocative visual and contextual enhancements that can enhance customer loyalty and brand awareness. Overall, augmented reality is an amazing example of technology that can make a positive contribution to various aspects of our lives by simplifying tasks, increasing efficiency, and turning our fantasies into reality. In the field of tourism, augmented reality allows tourists to have a unique and interactive travel experience and learn new facts about the history and culture of the places they visit. It can also be useful for museum staff and other professionals who can use augmented reality to create interactive exhibits and educational programs.

The use of augmented reality can lead to reduced work time and resources, increased efficiency, improved quality of life, and increased income. However, it is necessary to consider the limitations and risks associated with this technology and develop appropriate strategies to minimize these risks. Finally, one can note the possibility of developing dependence on technology, as is observed with many other technologies. Overall, augmented reality is a unique tool with enormous potential in various industries and areas of life. At the same time, it is necessary to consider the limitations associated with its use and pay due attention to the technology's ethical, technological, and other aspects.

References

[1] ITU, "J.301 Requirements for augmented reality smart television systems," 2015.

[2] G. Makamara and M. Adolph, "A survey of extended reality (XR) standards," in *2022 ITU Kaleidoscope- Extended reality – How to boost quality of experience and interoperability*, 2022, pp. 1–11.

[3] F. Arena, M. Collotta, G. Pau, and F. Termine, "An overview of augmented reality," *Computers*, vol. 11, no. 2, p. 28, 2022.

[4] "AR and VR Display Industry worth $8.2 billion by 2028," *MarketsandMarkets*. [Online]. Available: https://www.marketsandmarkets.com/PressReleases/ar-and-vr-display.asp. [Accessed: 15-Mar-2024].

[5] W. Wolfson, "Method Developed by Children's Hospital Los Angeles Researchers Automates Brain Cell Mapping," *Children's Hospital Los Angeles*. [Online]. Available: https://www.chla.org/blog/researchla/method-developed-children-s-hospital-los-angeles-researchers-automates-brain-cell. [Accessed: 20-Mar-2024].

[6] J. Kaur, "Virtual insights, real solutions: The promise of augmented reality in medicine," in *Approaches to Human-Centered AI in Healthcare*, Hershey, PA: IGI Global, 2024, pp. 20–41.
[7] D. Elford, S. J. Lancaster, and G. A. Jones, "Exploring the effect of augmented reality on cognitive load, attitude, spatial ability, and stereochemical perception," *J. Sci. Educ. Technol.*, vol. 31, no. 3, pp. 322–339, 2022.
[8] T. M. Wut and M.-L. P. Ng, "Virtual reality and augmented reality of tourism research: A review and research agenda," *J. Qual. Assur. Hosp. Tourism*, pp. 1–24, 2024.
[9] *AccuWeather*. [Online]. Available: https://www.tvbox.one/apps/accuweather.html#google_vignette. [Accessed: 01-Apr-2024].
[10] *National Geographic Explore VR Review*. [Online]. Available: https://www.worldofgeekstuff.com/national-geographic-explore-vr-review. [Accessed: 01-Apr-2024].
[11] ITU, "Y.1540Âă: Internet protocol data communication service - IP packet transfer and availability performance parameters," 2019.
[12] N. Ashtari, A. Bunt, J. McGrenere, M. Nebeling, and P. K. Chilana, "Creating augmented and virtual reality applications: Current practices, challenges, and opportunities," in *Proceedings of the 2020 CHI Conference on Human Factors in Computing Systems*, 2020.
[13] I. Wohlgenannt, A. Simons, and S. Stieglitz, "Virtual reality," *Bus. Inf. Syst. Eng.*, vol. 62, no. 5, pp. 455–461, 2020.
[14] D. Wu, Z. Yang, P. Zhang, R. Wang, B. Yang, and X. Ma, "Virtual-reality interpromotion technology for metaverse: A survey," *IEEE Internet Things J.*, vol. 10, no. 18, pp. 15788–15809, 2023.
[15] P. Smutny, "Learning with virtual reality: a market analysis of educational and training applications," *Interact. Learn. Environ.*, vol. 31, no. 10, pp. 6133–6146, 2023.
[16] B. Xie *et al.*, "A review on virtual reality skill training applications," *Front. Virtual Real.*, vol. 2, 2021.
[17] M. Javaid and A. Haleem, "Virtual reality applications toward medical field," *Clin. Epidemiol. Glob. Health*, vol. 8, no. 2, pp. 600–605, 2020.
[18] A. W. K. Yeung *et al.*, "Virtual and augmented reality applications in medicine: Analysis of the scientific literature," *J. Med. Internet Res.*, vol. 23, no. 2, p. e25499, 2021.
[19] H. Ghaednia *et al.*, "Augmented and virtual reality in spine surgery, current applications and future potentials," *Spine J.*, vol. 21, no. 10, pp. 1617–1625, 2021.

[20] A. J. Lungu, W. Swinkels, L. Claesen, P. Tu, J. Egger, and X. Chen, "A review on the applications of virtual reality, augmented reality and mixed reality in surgical simulation: an extension to different kinds of surgery," *Expert Rev. Med. Devices*, vol. 18, no. 1, pp. 47–62, 2021.

[21] T. L. T. da Silveira, P. G. L. Pinto, J. Murrugarra-Llerena, and C. R. Jung, "3D scene geometry estimation from 360° imagery: A survey," *ACM Comput. Surv.*, vol. 55, no. 4, pp. 1–39, 2023.

[22] S. Shin and S. Choi, "Hybrid framework for haptic texture modeling and rendering," *IEEE Access*, vol. 8, pp. 149825–149840, 2020.

[23] M. Scorpio, R. Laffi, A. Teimoorzadeh, and S. Sibilio, "Immersive virtual reality as a tool for lighting design: applications and opportunities," *J. Phys. Conf. Ser.*, vol. 2042, no. 1, p. 012125, 2021.

[24] A. Bellazzi *et al.*, "Virtual reality for assessing visual quality and lighting perception: A systematic review," *Build. Environ.*, vol. 209, no. 108674, p. 108674, 2022.

[25] C.-F. Chen and E. Suma Rosenberg, "Capture to rendering pipeline for generating dynamically relightable virtual objects with handheld RGB-D cameras," in *26th ACM Symposium on Virtual Reality Software and Technology*, 2020.

[26] B. Krajancich, P. Kellnhofer, and G. Wetzstein, "Optimizing depth perception in virtual and augmented reality through gaze-contingent stereo rendering," *ACM Trans. Graph.*, vol. 39, no. 6, pp. 1–10, 2020.

[27] I. Mahalil, A. M. Yusof, and N. Ibrahim, "A literature review on the usage of Technology Acceptance Model for analysing a virtual reality's cycling sport applications with enhanced realism fidelity," in *2020 8th International Conference on Information Technology and Multimedia (ICIMU)*, 2020, pp. 237–242.

[28] G. Gonçalves, H. Coelho, P. Monteiro, M. Melo, and M. Bessa, "Systematic review of comparative studies of the impact of realism in immersive virtual experiences," *ACM Comput. Surv.*, vol. 55, no. 6, pp. 1–36, 2023.

[29] C. Christou and A. Parker, "Visual realism and virtual reality: a psychological perspective," in *Simulated and virtual realities*, CRC Press, 2023, pp. 53–84.

[30] W. Xu, Z. Yang, D. W. K. Ng, M. Levorato, Y. C. Eldar, and M. Debbah, "Edge learning for B5G networks with distributed signal processing: Semantic communication, edge computing, and wireless sensing," *IEEE J. Sel. Top. Signal Process.*, vol. 17, no. 1, pp. 9–39, 2023.

[31] M. Alsabah *et al.*, "6G wireless communications networks: A comprehensive survey," *IEEE Access*, vol. 9, pp. 148191–148243, 2021.
[32] M. Banafaa *et al.*, "6G mobile communication technology: Requirements, targets, applications, challenges, advantages, and opportunities," *Alex. Eng. J.*, vol. 64, pp. 245–274, 2023.
[33] T. H. Chiang, S. J. Yang, and G. J. Hwang, "An augmented reality-based mobile learning system to improve students' learning achievements and motivations in natural science inquiry activities," *Journal of Educational Technology & Society*, vol. 17, no. 4, pp. 352–365, 2014.
[34] "Pioneering Surgical Navigation with Augmented Reality," *Novarad.* [Online]. Available: https://www.novarad.net/augmentedreality. [Accessed: 20-Mar-2024].

3

Holographic Images

3.1 Concept of Holographic Images

In the modern era of computer technology, visualization and 3D modeling are playing an increasingly important role. At the same time, holographic projection technology has recently gained momentum, which allows the projection of three-dimensional images using laser light. Holography is a multi-plane imaging technology that has already found wide application in various industries: from medicine to art. In this chapter, we will consider the features of holographic images and their capabilities in various fields of application.

The chapter analyzes the technologies for creating holographic images, considers methods of their storage and transmission, and also provides examples of existing applications of holography in medicine, education, science, industry, and art. The advantages and disadvantages of using holographic images will be outlined, and prospects for the development of this technology in the future will be presented. As a result, we will get a complete understanding of the capabilities of holography and its significance in the modern era of technology.

This topic is relevant due to the constant development of technology, where holographic images are one of the most promising areas. Their unique capabilities allow the creation of highly realistic images that can be used for education and entertainment and are also of great importance in scientific research and the development of new technologies. Moreover, this technology is actively used to create exciting light installations and exhibition displays. The purpose of this work is to gain a deeper understanding of the technology of holographic images and their capabilities, as well as to assess the prospects for their use in various fields of human activity. This work will be useful to those receiving higher education in the field of information technology, science, and art, as well as to everyone interested in this technology.

3.2 Applications of Holographic Images

Holographic imaging is an incredible technology that allows you to create 3D images that appear real. They are created by recording an interference pattern on a special plate or photographic film. To create these images, laser light is used, which creates patterns by refracting light waves on the surface of an object. One of the main advantages of holographic images is their detail and realism: they retain all the three-dimensional details and nuances that cannot be conveyed on a plane. This produces high-quality 3D images that can be used in medicine, education, science, industry, and the arts.

In medicine, for example, holographic images can be used to educate and train surgeons, as well as to create accurate three-dimensional models of patients' internal organs. In industry, holography allows the creation of realistic 3D models that can be used to design new products. Holographic images are also used in art to create interactive installations and exhibition displays. In general, holographic images are a current and promising technology that opens new opportunities in various fields of activity. They allow the creation of more accurate and detailed three-dimensional images, which makes them in demand in medicine, industry, and art.

In addition, holographic images have another important advantage – they can be used to transmit data. For example, using holographic crystals, it is possible to create high-capacity optical storage devices. This significantly increases the speed of transfer and storage of information compared to conventional CDs and flash drives. Also, holographic images can be used to create innovative security systems. For example, holographic plastic cards with images of unique three-dimensional structures and designs, such as complex engravings, can be used in various areas where a high level of authentication and information security is required.

Holographic images also open new possibilities in the field of entertainment, in particular for creating virtual reality. Real-time 3D imaging will enable the creation of highly realistic games and simulations that will be more interactive and engaging for users. As a result, holographic images have enormous potential for various fields of science and technology, from medicine and industry to entertainment and security. They are one of the most promising areas for the development of optical technologies, as evidenced by the increasing demand for holographic products and solutions in various fields.

Holographic images can also be used in scientific research. For example, they can be used to study physical processes in real time. This is relevant in the fields of physics, chemistry, and materials science, where it is necessary to observe processes at the molecular level and precisely control them [1].

Holographic images can also be used in the field of telecommunications. For example, they can be used to create very fast and secure data transfer channels between different devices. This can be especially useful in the areas of cloud computing and machine learning, where large amounts of information are required. Finally, holographic images can be used to create innovative products and services. For example, holographic projection devices can be used to create unique and fantastic shows and performances.

Holographic dissociative cards can also be used in the banking industry to protect client money from fraud. In general, holographic images open many possibilities in various fields of activity. They allow you to create more accurate, detailed, and realistic images, which makes them in demand in medicine, industry, art, science, and telecommunications. Moreover, holographic images can be used to create innovative products and services, making them one of the most promising technologies currently available.

In addition, holographic images can be used in medicine to create three-dimensional models of organs and tissues of the human body for diagnosis and surgical planning. Holographic images can also be used to train medical students and doctors. Holographic images can also be used in industry, for example, for testing and modeling various materials and products, which can significantly reduce the time and cost of their development and production.

Holographic images can also be used to create interactive user interfaces, making working with computers and other devices more convenient and efficient [2]. For example, holographic keyboards and control devices can reduce the time required to use them and reduce the likelihood of errors. Finally, holographic images can be used in art and design to create unique and innovative pieces. For example, holographic sculptures can be created using special materials and technologies. Overall, holographic imaging is a technology with enormous potential and greatly expands capabilities in various fields. They can improve accuracy, reliability, and safety of use, reduce development and production costs and time, and create new opportunities for creative work.

Another area where holographic images could have huge applications is in the entertainment industry. Holographic performances and concerts already exist, but technology allows us to create even more exciting and incredible shows that will leave an unforgettable experience for the audience.

Holographic images can also be used for virtual presence at events, conferences, and exhibitions [3]. This could be especially useful during the pandemic, when many events are moving to an online format, and holographic reproduction will allow users to feel almost on the spot. Another application of holographic images is in the gaming and virtual reality industries. They can provide an even more realistic and immersive gaming experience for players. Also, holographic images can be used for training in virtual piloting, surgery, and other areas of practical application.

In addition, holographic images have great potential for communication and social interaction. They can be used to create holographic video conferences, meetings, training, and other events that allow people to feel closer to each other, despite physical distance. Overall, holographic imaging is a technology that has enormous potential and a wide range of applications in various fields. They can provide opportunities for more accurate, realistic, and effective work and communication. The future of holographic technology looks bright and inspiring.

3.3 Applications of Holographic Images in Medicine

Holographic technologies are already used in medicine and have many potential applications. Some examples of existing applications of holography in medicine include the following [4, 5]:

1. **High-precision operations:** Holographic projection can be used to assist in high-precision operations such as neurosurgery and cardiac surgery. Surgeons can view a holographic model of a patient's body to better understand their anatomy and determine the best surgical strategy.
2. **Education and training:** Holographic models of organs and body parts can be used for teaching medical students and residents, as well as for training in surgery or invasive procedures.
3. **Diagnostics:** Holographic images can help identify diseases and disorders in the functioning of organs and tissues in the early stages. For example, holographic images can help detect tumors, lesions, and deformities in hidden areas of the body.
4. **Treatment:** Holographic technologies can be used to evaluate the effectiveness of treatment, educate patients on the use of medications, and monitor the progress of recovery after surgery or injury.
5. **Development of new drugs and medical treatments:** Holographic models can be used to develop and test new drugs and medical treatments, which can lead to more accurate predictions of their effects on the body and increased treatment effectiveness.

These are just some examples of the medical applications of holography that are already available today. With the advent of new technologies and developments, the number of applications of holography in medicine will only increase. Let us look at each of them in a little more detail.

3.3.1 High precision operations

The use of holography in precision surgery is one of the applications of holographic technology in medicine. It allows surgeons to obtain detailed three-dimensional images of a patient's organs and tissues in real time. This allows surgeons to more accurately understand the patient's anatomy and more accurately plan the operation, taking into account the individual characteristics of each specific case. Holographic projection can be used in various types of precision surgeries, ranging from brain surgery to organ transplants.

During surgery, holographic images can be projected onto a screen or special glasses so that the surgeon can see 3D images that, using a gesture control system, allow the image to be manipulated. When a surgeon uses holographic images, he can easily find the exact location of openings to access the patient's tissues and organs. This allows you to minimize the time spent in deep areas of the body and, therefore, reduce the risk for the patient.

In addition, the holographic projection allows the surgeon to visualize both his own instruments and selected points on the patient, which prevents the risk of infection due to continuous contact. The use of holographic technologies in high-precision operations in medicine is under constant development and improvement. Its use can reduce the risk of complications and increase the accuracy of the operation, which significantly improves the success of such procedures.

In addition, holographic projection can give the surgeon greater freedom when performing manipulations and operations, since he can more accurately see and understand all intersecting organs and vessels. This is especially important in complex operations, such as tumor removal, where damage to healthy tissue must be minimized.

Holographic images can also be used in the training of surgeons and medical students, allowing them to learn simulated virtual surgeries, which can reduce the costs of live research. Thus, the use of holography in precision medical operations is an important tool to improve the accuracy and efficiency of operations. This technology will continue to evolve and improve, opening new opportunities for surgeons and improving patient care. Examples of the use of holography in high-precision operations include the following [6, 7]:

1. **Brain surgery:** When performing brain surgery, holographic images can help the surgeon better understand the patient's anatomy and better plan the route to access deep brain structures.
2. **Tumor removal:** When removing tumors, holographic images can help locate the tumor, measure its size, and determine the route of access, thereby minimizing damage to healthy tissue.
3. **Cardiac surgery:** In heart surgery, holographic images can help the surgeon see the position of stents, vein placement, and other heart valve manipulations.
4. **Orthopedic surgery:** For joint replacements, holographic images can help determine the location of the prosthesis, the exact dimensions of the area where the prosthesis will be placed, and the best route of access.

One example of the real-life use of holographic technologies in tumor removal is the Philips Azurion system [8], which uses holographic projection for navigation during surgery. The Azurion system allows surgeons to project 3D images of the patient's body onto a special screen located in close proximity to the operating table. This allows surgeons to see the internal structures of organs and tumors in real-time and more easily navigate when removing a tumor.

Holographic projection is complementary to other navigation tools, such as ultrasound and magnetic resonance technologies, and can be especially useful for operations that require high precision and increased visualization of internal structures. Overall, the use of holographic technologies in tumor removal can help surgeons perform operations more accurately and efficiently, minimize risks to patients, and improve treatment outcomes.

An example of the real-life use of holographic technologies in cardiology is the EchoPixel True 3D system [9], which allows medical specialists to diagnose and treat cardiovascular diseases more accurately and effectively. An example of the real use of holographic technologies in orthopedic surgery is the Intellijoint-HIP system [10], which allows surgeons to perform hip replacements more accurately and efficiently.

The Intellijoint-HIP system uses holographic projection to create an accurate 3D map of the hip joint, allowing surgeons to pinpoint the location of bones and other structures before surgery. During surgery, the Intellijoint-HIP system helps monitor bone and joint activity in real time, helping surgeons ensure that the prosthesis is properly installed and functioning properly.

This minimizes risks for patients and reduces treatment and recovery time. In addition, the Intellijoint-HIP system provides surgeons with a wealth of patient information, including muscle function, joint coordinates, speed, and activity in real time. Overall, the use of holographic technology in orthopedic surgery allows for more accurate and efficient procedures, reduces risks for patients, and leads to faster recovery.

The EchoPixel True 3D system uses holographic projection to create three-dimensional images of the heart and blood vessels. Thanks to this, medical personnel can see all the structures of the heart and blood vessels in their natural size, including those previously hidden and inaccessible to other methods.

This allows cardiologists and surgeons to more accurately assess the condition of the heart and blood vessels, identify blood flow disorders, and detect possible problems, such as circulatory obstruction. In addition, the EchoPixel True 3D system allows complex surgical procedures, such as aortic repair or coronary artery bypass grafting, to be performed with greater precision and safety for patients. Overall, using holographic technology in cardiology can help improve diagnosis, treatment, and patient outcomes, as well as improve the quality of medical personnel in the field.

3.3.2 Education and training

Holographic technologies can be useful tools in the education and training of healthcare professionals, both students and professionals. Some examples of the use of holographic technologies in teaching and training in medicine include the following [11, 12]:

1. **Holographic models:** Instead of using traditional models and pictures, holographic models can provide a more accurate picture of medical cases. For example, instead of viewing a photograph of a patient's brain scan, a healthcare professional can use a holographic model of the brain to study its structure in more detail.
2. **Educational courses:** Holographic educational courses can help medical students better understand the anatomy and physiology of the body. Courses can contain varying difficulty levels, from basics to more advanced concepts.
3. **Surgical training:** Holographic technologies can train surgeons and residents in demanding surgeries, surgical techniques, and management of complex medical cases. For example, a holographic simulator could teach surgical practice to reduce the risk of errors and increase efficiency.

4. **Medical data visualization:** Holographic technologies can visualize medical data such as brain scans and other medical images, which can help better understand, diagnose, and analyze medical cases.
5. **New treatments:** Holographic technologies can help healthcare professionals develop and implement new treatments. For example, medical devices using holographic technology can help facilitate surgeries and provide quality medical care.

One example of the successful use of holography in training is True3D Viewer [13], a new generation of software for medical imaging. This software converts patient anatomical data into fully interactive 3D virtual reality images. These innovations allow medical holography to be used for healthcare research, hospital teaching, and medical education. Last June, one of Minnesota's largest health systems, CentraCare, successfully completed the world's first structural heart surgery using this technology. "EchoPixel's True3D pre-planning software has helped us reduce procedure times by more than 27% and improve optimal outcomes by 20%," said Jacob Dutcher, MD, interventional cardiologist and director of the Structural Heart Program at the Heart and Vascular Center, CentraCare, which performed surgery on a patient with atrial fibrillation.

One example of the real-world use of holographic technologies in medical data visualization is the Microsoft HoloLens system [14], which allows medical professionals to view medical data in the form of holographic images. The Microsoft HoloLens system uses holographic technology to create 3D images of medical data in real time. This allows medical professionals to view and analyze data without having to interrupt operations, which can improve the accuracy and efficiency of medical procedures.

Using Microsoft HoloLens, healthcare professionals can view 3D images of X-rays, MRIs, and CT scans and analyze models of organs and other structures in 3D space. This allows healthcare professionals to obtain more accurate information about a patient's condition and develop more effective treatment plans. In addition, the Microsoft HoloLens system can also be used to train medical students, allowing them to learn anatomy and organ functionality more effectively, and practice medical procedures without the need for real patients.

3.3.3 Diagnostics

In diagnostics, holographic technologies are used, for example, through the use of systems such as HoloAnatomy [15], which allows medical specialists to study and diagnose various pathologies in real time. The system uses holographic technology to create a 3D patient anatomy model. Medical professionals can then view this model through holographic glasses and view it in real time. This means that medical professionals can use this system to diagnose various pathologies such as tumors, cysts, and inflammatory processes.

This allows healthcare professionals to obtain more accurate information about a patient's condition and develop more effective treatment plans. Additionally, it can be used to teach medical students, allowing them to effectively learn anatomy and organ functionality and practice medical procedures without the need for real patients.

The HoloAnatomy system is a unique medical instrument based on holographic technologies. It is designed to facilitate and improve the training process for medical personnel, providing them with access to three-dimensional visualization of internal organs and tissues.

It uses real-life data collected through MRI, CT, and other scanning techniques to create an accurate 3D model of a patient's anatomy. This model can be viewed in real time using holographic glasses, allowing medical professionals to gain a deep understanding of the patient's internal organs and tissues. The use of holographic glasses allows medical personnel to obtain additional data about certain patient diseases in real time, which can facilitate the diagnostic process and increase the effectiveness of treatment.

In addition, the system allows medical staff to conduct virtual training and simulations of medical procedures, which can help improve professionalism, reduce patient exposure, and improve procedural accuracy. The HoloAnatomy system can be useful not only in the education of medical personnel but also in other areas such as scientific research and planning of medical procedures. This system allows medical personnel to obtain more accurate and complete information about internal organs and tissues, which can significantly improve treatment results.

3.4 Applications of Holographic Images in Science

Holographic images are used in many fields of science, including technology and astronomy. Holography allows one to obtain more accurate and detailed images of three-dimensional objects than is possible using conventional two-dimensional images. Examples of the use of holographic images in science include the following:

1. **Capture technical documentation:** Some objects may be too complex to follow in 2D space. Holography produces images that can help researchers better understand these objects.
2. **Visualization in medicine:** Holographic images are used in medicine to create three-dimensional models of internal organs and tissues. This facilitates the design of implants and other medical devices and helps to visualize various pathological processes.
3. **Space exploration:** Holographic images are used in astronomy to obtain more detailed information about space objects such as exoplanets.
4. **Presentation materials:** Holographic images can be used to create interactive presentations in science centers and museums, showing the public 3D models of tissues, biological objects, and more. The general idea is that holographic images create a more accurate and detailed view of three-dimensional objects, which can be useful for solving scientific problems in many different fields.

One example of the use of holographic technologies in astronomy is the use of NASA's Kepler space telescope [16]. It has been used to study the characteristics and motion of exoplanets. For this purpose, special holographic optics were developed, which makes it possible to create images of space objects detected by the telescope. Discoveries made with the Kepler Space Telescope include the discovery of the first Earth-like planet in the habitable zone of another star and more than a thousand exoplanets in the Milky Way galaxy alone. The use of holographic optics in this mission improved the quality of images and made more detailed and accurate measurements of the characteristics of space objects, which will be useful for further exploration in space.

NASA has been using holography to create 3D models of space objects for many years. In 2016, a series of experiments were conducted that showed the high efficiency of this method. In particular, NASA used the holographic method to create 3D models of unique meteorite samples that were found in Antarctica using the Astromaterials 3D Explorer application [17]. To do this, agency scientists studied spectral data obtained by an infrared camera, which was used to reconstruct the three-dimensional structure of meteorites using a holographic process.

The essence of the process is that the light source object is divided into two ray waves – the reference and the object. The object wave passes through the object and then overlaps with the reference wave. This results in the creation of an interference pattern called a hologram, which contains information about the shape and location of the object. To create a three-dimensional model of the meteorite, laser light was used to scan the meteorite and create a hologram, and then data processing was carried out, which made it possible to create a detailed three-dimensional model of the object. The created three-dimensional model of the meteorite showed scientists its structure and shape, which can be useful in studying the origin and composition of meteorites.

In general, the holographic method of creating 3D models is a powerful tool for studying space objects, as it allows you to obtain detailed information about their shape, structure, and composition.

3.5 Applications of Holographic Images in Industry

Holographic images have found wide application in various fields, including industry and art. Examples of the use of holography in industry include the following [18, 19]:

- Holographic stickers on holograms on goods to protect against counterfeiting.
- Holographic sensors for monitoring rotation and deformation of equipment.
- Holograms for quality control of welding and soldering.
- Holographic devices for determining the concentration of various components in chemical mixtures.

Examples of using holography in art include the following [20]:

- Use of holographic images in sculpture to create three-dimensional compositions.
- Creation of 3D holographic paintings and panels.
- Use of holographic images to create interactive installations.

Holographic images can be used to create dramatic effects, add dimension and depth to images, and create a unique visual experience for the viewer. All this suggests that holography is widely used in industry and art, making it possible to use it to protect against counterfeiting, control production quality, and create three-dimensional compositions.

3.5.1 Holography in protection against counterfeiting

Holography is one of the most effective anti-counterfeiting methods used in the industry. Holographic images are unique because they are created by the interference of light on a surface, the shape and geometry of which is unique to each hologram. This allows you to create holograms that are very difficult to fake. The hologram can be in different places on the banknote, for example, on the foil, watermarks, or the paper itself. The hologram on the banknote contains complex images and texts that can only be seen from a certain angle. This makes it easy to verify the authenticity of a banknote without special equipment.

Holographic images can be used in various documents, including auto reports. Unique holograms can be created using laser engraving, which creates a special texture on the surface that can be used as an additional security feature against counterfeiting. Holographic images can be used to create unique beacons embedded in trademarks and logos to protect against counterfeiting. Holograms may contain information that trademark owners can only verify using special equipment.

Holographic images are used in various fields, including industrial, banking, and automotive industries, for anti-counterfeiting purposes. Holograms can be used as an additional means of security for various objects such as banknotes, documents, and trademarks and can help to verify the authenticity of these objects easily.

In addition, holographic images can also be used to create unique packaging designs for storing various products. They can be combined with signatures, colors, fonts, and images, and together with bag label technology, create anti-counterfeit protection that is reliable and quite effective. Another interesting example of the use of holography in protection against counterfeiting is the latest types of financial cards. These credit cards use holographic protection as a method of preventing fraud. Often, these cards contain a hologram on the front, which may include elements such as the owner's photo, fingerprints, bank logos, name, and line of credit.

Integrating holograms onto new types of financial cards is the latest way to use holography in anti-counterfeit protection. For example, some credit cards now contain holograms on their surface. These holograms may contain unique elements such as photographs, logos, bar codes, and other text that can only be seen from certain viewing angles. This provides additional protection against fraud and identity theft for credit card holders. Holograms can also contain special elements to identify the user biometrically, such as fingerprints or scanning a barcode on a card, increasing security even further.

There are also different variations of holograms for the creative use of holography as a design element and card protection. For example, in 2020, Mastercard released a camouflage credit card for the first time, which had a mirror surface on a holographic insert capable of creating optical illusion effects. In addition, the image on the hologram changes depending on the viewing angle, making the card more reliable and less susceptible to counterfeiting.

3.6 Summary

Holographic images are a unique object visualization technology that allows you to get a more detailed and complete picture of a three-dimensional object than is possible with conventional images. They are widely used in the modern world in various fields such as medicine, science, industry, art, and entertainment. In medicine, holographic images are used to create three-dimensional models of organs and tissues, which allows surgeons to more accurately plan operations and determine the optimal route for surgery. In science, holographic images are used to create three-dimensional models of molecules and atoms, which makes it possible to study their structure and properties.

One of the promising areas for using holographic images is virtual and augmented realities. Holographic technology can be used to create more realistic visual environments, allowing for more interactive and immersive experiences for the user. In business, holographic images can be used to create virtual stores and showrooms, allowing customers to view and test products in real time, without physically being in the store. Additionally, in education, holographic images can be used to create interactive lessons and simulations to help students better understand and remember course material. However, holographic images also have limitations and raise certain technical and ethical issues related to copyright, privacy, and data security. For this reason, before using holographic images in any field of activity, it is necessary to consider and evaluate all the risks and problems associated with their use.

Although holographic imaging is a potentially promising technology, there are some limitations and risks associated with its use. One of the main risks associated with the use of holographic images is violation of copyright and intellectual property rights. Creating holographic images from someone else's material may result in legal liability, especially in cases where permission to use has not been requested. In addition, using holographic images with sensitive information may lead to data leakage and privacy violations.

The potential use of such images in medicine, including the use of holographic reconstructions of patients, may raise ethical and privacy concerns. It is also worth noting that the use of holographic images requires impressive technical resources and knowledge. Available software for creating holographic images is still not at a high level, making the development process long and labor-intensive.

Overall, despite some limitations and risks, holographic images still represent a technological advancement that can be used to create more innovative products and services in a variety of industries, from industrial to creative fields.

While holographic imaging comes with some limitations and risks, it also represents a potentially powerful tool for a variety of industries and applications. Here are some examples of the positive impact of using holographic images:

1. **Gaming industry:** Game developers can use holographic images to create a new level of interactivity and enhance the gaming experience. Such images can help create a more realistic gaming environment that allows players to become more immersed in the game.
2. **Medical industry:** The use of holographic images can significantly improve the quality of medical diagnosis and surgical planning. Doctors will be able to see the body's internal structure in three dimensions, making it easier to understand the location of organs and nerves and help choose the best surgical path.
3. **Advertising:** Holographic images can create more dynamic and attractive advertisements. They can help grab the attention of potential customers by showing products in a new light and attracting attention with eye-catching effects.
4. **Education:** Holographic images can help improve learning in various disciplines, from literature to science and technology. They can help students see things more realistically and interactively, making complex concepts easier to understand and increasing interest in education.

Overall, using holographic images can help create new capabilities and improve processes in various industries and applications. However, it is necessary to consider the limitations and risks associated with this technology and develop appropriate strategies to minimize these risks. In addition, using holographic images also requires certain requirements from communication networks. These requirements include the following:

1. **Broadband internet:** Holographic images require large bandwidth to transmit images and videos with stereo sound. Therefore, for high-quality viewing and transmission of holographic images, a broadband internet connection is required.

2. **Low latency:** Holographic images require low latency to create the illusion of three-dimensional space. High latency transmission systems may cause irregular images and visual distortion.
3. **Network connection reliability and stability:** Unpleasant surprises caused by network connection interruptions can cause problems when viewing holographic images. Therefore, it is necessary to ensure the reliability and stability of the network connection to meet the necessary quality requirements.
4. **Device compatibility:** Different devices support different holographic image formats and technologies. Therefore, it is important to select a holographic imaging technology that is compatible with the user's target devices.

Thus, using holographic images requires a high-quality and stable network infrastructure compatible with the target user devices to ensure high-quality holographic displays. So far, the current network infrastructure does not fully meet the requirements for using holographic images, as many users are experiencing problems with the stability and speed of internet connections. Holographic images require high bandwidth and low latency and highly reliable and stable network connections.

Although networks with higher speeds and capacity, such as 5G, exist, many areas still do not have access to this technology, and some countries are just beginning to implement it. In addition, many users face problems with network connection speed and stability when using current cellular and wired internet technologies.

Thus, for holographic images to become more accessible to a wider audience, further development of network communication infrastructure and technological solutions are required to solve the problems of speed, reliability, and stability of internet communications. However, despite these limitations, holographic images still present great potential for use in various fields such as medicine, education, entertainment, architecture, and many others.

References

[1] M. Javaid, A. Haleem, and I. Khan, "Holography applications toward medical field: An overview," *Indian J. Radiol. Imaging*, vol. 30, no. 3, p. 354, 2020.

[2] T. Tahara, X. Quan, R. Otani, Y. Takaki, and O. Matoba, "Digital holography and its multidimensional imaging applications: a review," *Microscopy (Oxf.)*, vol. 67, no. 2, pp. 55–67, 2018.

[3] W. Haider, M. A. Abbasi, R. Shams, S. Khan, and S. S. A. Jafri, "Towards representation of real entities using holographic technology," *JAET*, vol. 6, no. 2, pp. 91–101, 2022.

[4] C. Gsaxner *et al.*, "The HoloLens in medicine: A systematic review and taxonomy," *Med. Image Anal.*, vol. 85, no. 102757, p. 102757, 2023.

[5] L. Wang, Y. Na, and G. Zhu, "Application of holographic imaging in partial nephrectomy: a literature review," *Holist Integ Oncol*, vol. 3, no. 1, 2024.

[6] Y. Saito *et al.*, "Intraoperative 3D hologram support with mixed reality techniques in liver surgery," *Ann. Surg.*, vol. 271, no. 1, pp. e4–e7, 2020.

[7] M. Jeyaraman, N. Jeyaraman, S. Ramasubramanian, A. Nallakumarasamy, and A. Shyam, "Revolutionizing Orthopedic Surgery: The Integration of Holographic Technology," *Journal of Orthopaedic Case Reports*, vol. 14, no. 3, 2024.

[8] "The next generation image-guided therapy platform," *Philips*. [Online]. Available: https://www.philips.sa/en/healthcare/resources/landing/azurion?_ga=2.181964515.574990644.1713590718-1285984430.1713590717. [Accessed: 04-Jan-2024].

[9] "TRUE3D - SURGICAL PLANNING." [Online]. Available: https://echopixeltech.com/true3d. [Accessed: 02-Apr-2024].

[10] "Designed for Surgeons — Intelligent Hip Navigation." [Online]. Available: https://www.design-industry.com.au/navbit-sprint. [Accessed: 05-Apr-2024].

[11] M. N. Hassan Ja'ashan, H. Abdulaziz Alfadda, and H. Saleh Mahdi, "Using a holographic application in learning medical terminology for English as a foreign language students," *Interact. Learn. Environ.*, vol. 32, no. 2, pp. 600–613, 2024.

[12] H. Türk and M. Seçkin Kapucu, "Innovative technology applications in science education: Digital holography," *J. Educ. Environ. Sci. Health*, 2020.

[13] "Hologram for the kings of medicine: the present and future of holography in healthcare," *EverCare*. [Online]. Available: https://evercare.ru/news/gologramma-dlya-koroley-mediciny-nastoyaschee-i-buduschee-golografii-v-zdravookhranenii. [Accessed: 15-Mar-2024].

[14] "Microsoft HoloLens," *Wikipedia*. [Online]. Available: https://en.wikipedia.org/wiki/Microsoft_HoloLens. [Accessed: 01-Apr-2024].

[15] "Microsoft's HoloAnatomy app for learning anatomy," *AR Soft*. [Online]. Available: https://vc.ru/u/703859-ar-soft/328073-prilozhenie-holoanatomy-ot-microsoft-dlya-izucheniya-anatomii. [Accessed: 20-Mar-2024].

[16] "NASA's Kepler space telescope," *NASA*. [Online]. Available: https://science.nasa.gov/mission/kepler/. [Accessed: 15-Feb-2024].

[17] "Every Rock Tells a Story," *NASA*. [Online]. Available: https://ares.jsc.nasa.gov/astromaterials3d/sample-details.htm?sample=ALH84001-55. [Accessed: 25-Jan-2024].

[18] S. Barile, C. Simone, F. Iandolo, and A. Laudando, "Platform-based innovation ecosystems: Entering new markets through holographic strategies," *Ind. Mark. Manag.*, vol. 105, pp. 467–477, 2022.

[19] M. Javaid, A. Haleem, R. Vaishya, S. Bahl, R. Suman, and A. Vaish, "Industry 4.0 technologies and their applications in fighting COVID-19 pandemic," *Diabetes Metab. Syndr.*, vol. 14, no. 4, pp. 419–422, 2020.

[20] M. Hernandez-de-Menendez, C. Escobar Díaz, and R. Morales-Menendez, "Technologies for the future of learning: state of the art," *Int. J. Interact. Des. Manuf. (IJIDeM)*, vol. 14, no. 2, pp. 683–695, 2020.

4

Holographic-type Communications

4.1 History of Hologram Development and Technology Features

Current research aims at rethinking communication potentials. Perspectives for building and implementing communication innovations based on holographic communication have been created and defined. A hologram is an optical clone, or, in other words, a recording of a light field that preserves the original depth and parallax of real 3D objects (people/objects). Holography is a technology used to record and create holograms. Holography was developed by Dennis Gabor in 1948, who received the Nobel Prize in Physics in 1971 [1]. Since then, various technologies for the creation and display of holographic images have been embodied in life. Unlike video and images projected on 2D screens, a hologram requires a significantly larger amount of data to fully demonstrate a true 3D structure with the additional properties of depth and parallax. As a rule, holograms require data transfer speeds of several Gbit/s or even Tbit/s, depending on the size and resolution of the image. As a result, hyper-real holograms are usually generated and (partially) displayed in the same place without any remote transmission due to limited network bandwidth [2].

Holographic-type communication (HTC) allows the transmission of holograms and other multisensory media (Mulsemedia) via wireless and wired networks to remote locations. The HTC system basically consists of three parts, namely the source, the communication network, and the receiver of the data. Holograms and multimedia are created or stored at the source, forwarded through HTS networks, and then processed and presented at the destination [3].

Hologram transmission has been studied and implemented for more than 10 years in some ideal scenarios with limited QoE. In 2008, Cisco demonstrated the On-Stage Telepresence Experience, where a user from India virtually met two users from the USA [4]. Unlike videoconferences, where remote users are displayed on a 2D (two-dimensional) screen, the holographic system displays two remote users on a 3D (three-dimensional) display as life-size holograms. Having sufficient resolution, the holograms looked quite realistic, which allowed the system to provide a good immersive experience. The Telehuman project [4] (Telehuman in 2012 and Telehuman2 in 2018) demonstrated inexpensive constructions of HTC using cylindrical holographic displays. Telehuman2 can provide 10 frames per second (fps), and the delay from capture to projection is about 200 ms. Also, instead of advanced displays with a light field, head-mounted displays (HMD) of augmented reality (XR) [5], for example, Microsoft HoloLens, are used to display holograms. Developed in 2015, LiveScan3D can display holograms in the same room as the user.

Experiments are also carried out, aimed at revealing the limit values of the reception–transmission time (e.g., round trip time (RTT)) for holographic traffic. Thus, in [6], three different values of the distance between common points are tested, in which the source is located in Guildford, Great Britain, and the three destination points are located in London, United Kingdom (the signal transit time there and back is about 4 ms), Virginia, USA (RTT is about 85 ms), and Seoul, South Korea (RTT is about 285 ms). Based on the obtained data, it was revealed that the user in London could stably receive 30 frames per second, while the subscriber in Seoul experienced a decrease in performance.

Recently, holographic technologies have been widely used in various entertainment events, such as the Whitney Houston Hologram Tour in 2020 [7] and the AVBA Voyage Concert in 2022 [8]. Even though in the last decade there has been a rapid explosion of HTC, the worldwide development of existing prototypes is held back by low frame rates and insufficient resolution of three-dimensional images. The transfer of holographic streams in existing communication networks is an urgent problem due to the large size of the data and the strict requirements for delay. The development of wireless and wired networks of the next generation, such as 6G and above, as well as the use of artificial intelligence (AI)/machine learning (ML) in networks and telecommunications, will make it possible to provide high-quality hyper-realistic HTC at any distance with the required levels of quality of service (QoS) and quality of experience (QoE). It is expected that HTC will become an integral communication option in Networking 2030, i.e., IMT2030/6G [9].

4.2 Creation and Display of Holograms

Computer-generated holograms can be divided into two categories: image-based solutions and volume-based solutions [10]. Image-based solutions, such as video with a light field, use a large number of frames at different angles shot by arrays of cameras. The spatial angular interval between the cameras determines the resolution. Volume-based solutions, such as point clouds, represent real 3D objects using volume-based 3D pixels. Holograms differ from typical 3D images in that for the perception of ordinary 3D video it is necessary to use special glasses that create 3D effects. Holograms, in turn, can be observed with the naked eye.

Arrays of cameras are used to form images of 3D objects at different angles. These images are then processed based on depth and parallax information. In the conducted research, special attention will be paid to the method of working with a cloud of points, due to its high efficiency in the representation of 3D objects in relation to the holograms of the light field, using a large number of combined images. The point cloud is relatively small and can be effectively processed and transmitted, which has determined the main trend of holographic representation [11].

Basically, there are three types of displays for holographic communication that can be introduced as follows [12].

1) **Head-mounted XR displays:** Head-mounted displays (HMDs) are widely used in extended reality (XR), which combines augmented reality (AR), mixed reality (MR), and virtual reality (VR). The presented technologies form two different views for the left and right eyes (parallax), thereby providing an effect similar to that which a person receives when perceiving the world in real life. In this implementation, users have to use XR HMD to observe holograms, which imposes some restrictions. Head-mounted displays support a strictly fixed number of viewing angles and often cause extreme fatigue in users after prolonged use. However, since the display is located close to the eyes, the amount of information that needs to be transmitted to the HMD is limited, and the required bandwidth is low compared to light-based displays.
2) **Multi-view volumetric displays:** They can support multiple viewing angles without the need for any glasses or HMD. However, their resources in terms of the number of possible viewing angles are still limited compared to light-field displays. These are relatively small holographic displays that can be used for mobile devices and monitors. They can typically only support a single user, and eye-tracking technologies

can be used to adaptively render holographic content. In the current study, the holographic frustum falls under this category of displays.

3) **Light-field displays:** They can provide thousands of different viewing angles and simultaneously broadcast information to multiple users [13]. No external glasses or HMDs are required. A two-dimensional hogel array is used to display the light field, and each hogel (part of the light-field hologram, a small holographic optical element) can generate unique light intensity in different directions to recreate the light field [14].

Unlike 2D squeaks, hogels are controlled using a vector that includes the hogel's position in a 2D array and a vector describing the direction and intensity of the emitted light for different viewing angles. Like antenna arrays, hogels are arranged in a grid structure to generate light fields and create holograms. In reality, when a user observes an object, they see a light field reflected from it. The displays under consideration work in a similar way, presented in the form of an array of hogels and forming a holographic image for free viewing in 3D without auxiliary devices based on the physical properties of light fields. Synthetic hogels are typically rendered using double frustum, oblique slices, and cubes or polygon/voxel tracing/ray casting.

Light field displays are expected to be widely used in future HTC systems as they can deliver hyper-realistic user experiences. The required data rate of a light field display is directly dependent on its size, hogel intensity, and directional resolution (D_T). The directional resolution is determined by the size of the hogel, and for example, for a square hogel, $D_T{}^2$ represents the number of rays that can be generated. Thus, for a display of 1.5 m $\times$ 0.75 m with 0.5 mm hogels and D_T = 128, which can be used to display people, the data size for one frame is calculated as follows [15].

$$B = \frac{1.5\ \times 0.75}{0.0005^2} \times 128^2 \times 3 \approx 221 \text{ GB}, \tag{4.1}$$

where we assume that the number of RGB (red, green, blue) bytes per pixel is 3. Assuming that such a display will be used for the HTC service in real time, considering the refresh rate of 30 frames per second, the required data transfer rate will be 6.63 Tbit/s. Such high data transfer rates are beyond the capabilities of modern telecommunications infrastructure, which confirms the need to explore new designs of fundamental communication systems and software to support HTC.

4.3 Multimedia and Holographic-type Communication

Simply sending a hologram from a source to a virtual or remote destination cannot fully provide an immersive experience. To achieve a result as close to reality as possible, HTC will use all five senses of human perception, including vision, hearing, touch, smell, and taste. For example, a user at a remote location will be able to be in the same environment as a user at the source location, i.e., the users will be in the same space [15].

An environment that includes three or more human senses is called multimedia and is created using various sensors at the source side and then presented using actuators at the destination side. To achieve a truly immersive effect, all available human senses must be engaged. Because of this, in addition to recreating factors that simulate the effects on the main sensory organs, it is possible to further improve the user experience by using the senses of balance and location and supporting skin sensations (temperature, wind, humidity, etc.).

Most multimedia technologies appeal to only one or two human senses. Recently, XR and haptic communication have begun to actively embrace touch to develop more interactive and immersive applications. Although some XR devices support smell and taste, they are not widely used. In addition, XR, in conjunction with HMDs, provides an integrated platform for various sensors and actuators and allows the creation and presentation of multimedia functionality. In HTC systems without HMDs, it is necessary to use external sensors and manipulate devices, which significantly complicates the task.

Multimedia is a set of information that can simultaneously influence three or more senses [16]. Multimedia tools, aimed at working with video, audio, images, etc., use cameras and microphones to collect data at the source. At the destination, videos and images can be displayed on monitors, and audio files can be played through speakers. Likewise, multimedia systems use a variety of touch sensors at the source to collect data and a variety of actuators, manipulation devices, displays, and speakers at the destination to recreate a similar environment.

Recording sensors are placed in various places, with mandatory requirements to ensure synchronization of the collected data. Otherwise, the user at the destination may experience inconsistent and unnatural sensations. Once the receiving end receives the media traffic, it processes and renders it. HTC systems may not involve all senses due to the lack of sensors or actuators. Users or applications can select which perception tools to use, depending on available hardware and privacy and security considerations.

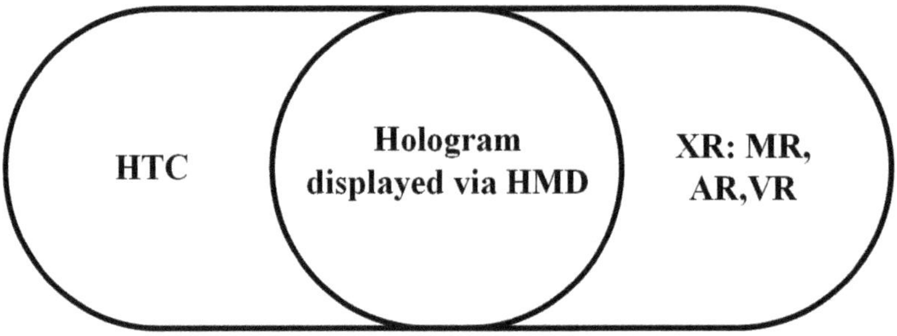

Figure 4.1 The connection between HTC communication and extended reality (XR). The intersection between XR and HTC is a hologram displayed using XR HMDs [15].

HTC is a unique technology, different from XR. Figure 4.1 presents the relation between HTC and XR. XR users can use HMD to access HTC content, but augmented reality is not the only or primary option for receiving media in holographic communication. Users without an HMD can observe and view HMD content with the naked eye and on light-field displays. Thus, HTC and XR have intersection points, but each has its own characteristics.

4.4 Relevance of HTC Development

It is expected that within 10 years, existing methods of remote interaction between people will become obsolete, as new forms of communication will emerge that will lead to full, immersive immersion in an artificially created (remote) environment. Of course, this will not happen overnight, and therefore, it is necessary to begin to formulate an idea of the possibilities of implementing holographic communication services and determine the requirements they will impose on the characteristics of data transmission networks. So, thanks to the significant development of holographic display technology in recent years, such as HoloLens from Microsoft, its use will become a reality in the next decade. For example, holographic telepresence will allow remote participants to be projected as holograms into a conference room or enable online training participants to interact with ultra-realistic objects.

The relevance of the research topic is also due to the rapid development of holographic communication technologies and the resulting need to determine the necessary network characteristics of lines, the performance of control, routing, switching, and user devices for the widespread implementation of these services. Currently, projects using holographic call technologies and hologram visualization are being actively implemented, and there are trends toward an increase in holographic content [17].

Based on qualitative and quantitative market research, it is clear that for holographic communication to become an accepted form of communication, it will require:

- flexible, scalable, and fault-tolerant platform for technology support,
- new and socially acceptable XR devices (e.g., glasses), and
- a simple, attractive, reliable solution accessible and intuitive for everyone.

It is possible to fully implement holographic communication projects only through modern holographic network applications using new-generation networks and high-performance information processing methods. With the widespread of holographic services, it will be necessary to ensure the transfer of extremely large volumes of data containing generated holograms with the required QoS indicator.

The researchers explain this by saying that holographic network application traffic is extremely sensitive to data transmission delays. Holographic communication requires high throughput and low latency, the maximum value of which should not exceed 1 ms. The term "data transmission delay" in this work is interpreted as a certain period of time that can be recorded from the moment a data packet is transmitted until the moment it is received. The delay in encoding information can be reduced by improving algorithms for processing and converting holographic copies. The delay introduced by the data network can be corrected by developing a new network architecture with an optimal infrastructure.

The main indicators of quality of service are network throughput and data transmission delay. Thus, the 3GPP standards provide requirements for network application delays: video calls – 150 ms, VoIP – 100 ms, and real-time video games – 50 ms. In addition, it is necessary to highlight such characteristics as network availability and reliability, which can be assessed by monitoring the level of service over a certain period of time or by determining the utilization rate. Recently, the quality of perception metric QoE has become increasingly relevant, complementing the already existing QoS indicator and determining the subjective assessment of users, such as, for example, the quality of the transmitted video.

From a civilizational perspective, the holographic method of recording and reconstructing the wave field, which applies to waves of various natures, will develop exponentially. At this stage, developments are actively underway to use optical holography technology in medicine, the film industry, and business – to provide various services. The need for distance learning necessitates the organization of this process at a new level, the level of holographic communication with the psychophysiological effect of the presence of a teacher.

Accordingly, research in this area will focus on studying increasingly complex types of traffic, developing innovative applications and technologies for creating holographic copies using multiple types of data, and addressing the problems of transmitting and optimizing holographic images. A group of scientists (S. Schwarzmann, P. Breitbach, T. Zinner, and M. Rost) took an interesting approach to consider the influence of various parameters on the performance indicators of holographic technology. They characterize the problem as follows: since there is still no holistic methodology for assessing the perceptual QoE of holographic models, we have only a limited understanding of the impact of various parameters on the performance metrics of adaptive 3D video streaming under different operating conditions. In this context, parameters include internal influencing factors such as the number of quality levels, quality switching thresholds, or segment duration, while operational settings refer to external influencing factors such as network characteristics or video content [18].

In [19], the author describes a "high-fidelity holographic society" in which holographic presence will allow remote users to be in a visualized space. For example, these could be technicians performing remote repairs and troubleshooting, doctors performing remote surgeries, or lecturers providing distance learning. The authors noted that 4G and the expected data transfer rates of 5G networks might not allow the implementation of the above technologies, but 6G will be able to fully cope with the transmission of traffic recorded from multiple viewpoints, from different angles, and positions of the observer of a holographic object.

Another promising possibility highlighted in [20] is the haptic internet, where different sensory experiences could be combined with holograms. For example, the use of holograms as a means of communication in mobile devices, together with the function of recognizing emotions, will allow us to monitor our mental health and promote social communication.

5G networks will provide a wide range of services, including enhanced mobile broadband (eMBB), ultra-reliable low latency communication (uRLLC), and massive machine-type communications (mMTC) [21]. However, the volume of wireless data transmission and the number of connected devices is expected to increase hundreds of times per given cubic meter.

Additionally, information-intensive applications, such as sending holographic videos, require bandwidth that is not currently available in millimeter waves. This situation poses challenges in terms of range and spatial–spectral efficiency, as well as the required frequency spectrum bands for communication. Therefore, a wider bandwidth of the radio frequency spectrum is a necessity and can only be obtained in the sub-terahertz and THz ranges.

Moreover, the recent surge in a variety of mobile applications, especially those supported by artificial intelligence (AI) technology, is generating heated discussions about the future evolution of wireless communications [22]. These challenges have prompted industry and academia to begin developing the concept of wireless communication systems. The upcoming 6G aims to provide communication services for the needs of the 2030s and maintain the sustainability and competitiveness of wireless communication systems.

In addition to typical 5G use cases, there are three advanced schemes considered in 6G: ubiquitous mobile broadband (uMBB), massive ultra-reliable low-latency communication (mULC), and ultra-reliable low-latency broadband communication (ULBC). Figure 4.2 presents the relation between such use cases. Ultra-reliable low-latency broadband communication supports applications that require not only uRLLC but also extremely high bandwidth, such as immersive games on HTC [23]. This scenario is expected to benefit use cases such as HTC, XR, haptic internet, multi-sensory experiences (affecting multiple senses), and pervasive intelligence (the ubiquity of smart devices and machines powered by artificial intelligence).

Holographic-type communication will take human communication to the next level: it will provide immersive 3D interaction in real time, which can be carried out even using consumer-grade mobile devices. Thanks to technology that can provide users with full face-to-face interaction, the subject matter opens up many opportunities in areas such as manufacturing, telecommunications, education, medicine, etc. Holographic communication will be of great importance for remote communication formats, as it will overcome the feeling of distance from colleagues, friends, and family. Growing awareness of the benefits of holographic communication is fueling the rapid growth of its use in various fields, from medical imaging and telepresence to remote collaboration and co-creation.

HTC is still under active research and development and, therefore, is not yet widely used by the masses. However, some companies are already developing and deploying holographic technologies in various fields, and the best prospects can be classified into five categories, namely: holographic telepresence and conferencing, education and training, healthcare, entertainment, and

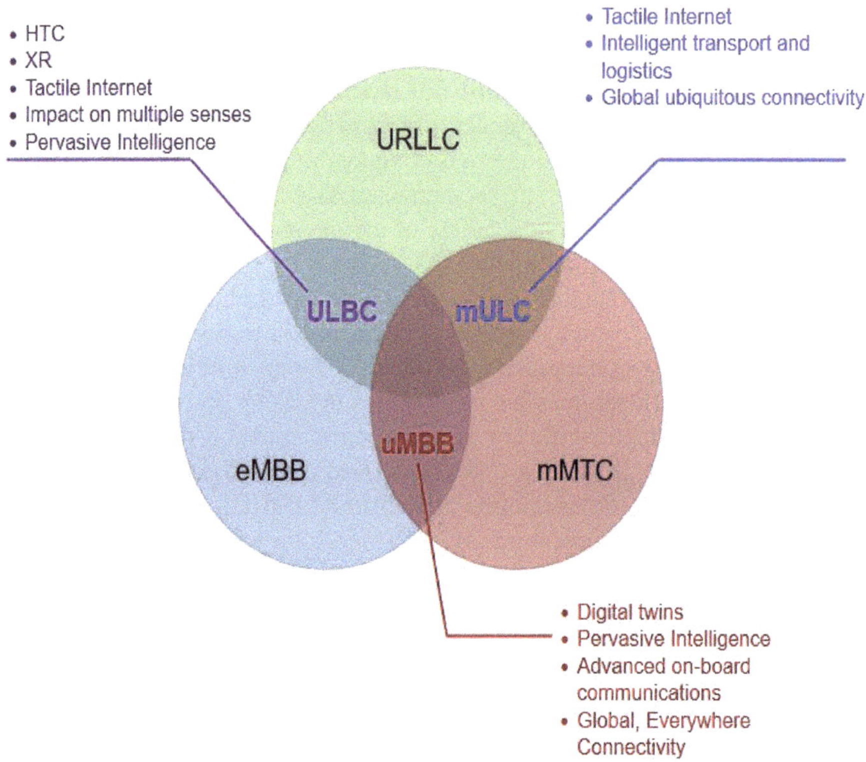

Figure 4.2 Three advanced 6G system scenarios, namely uMBB, ULBC, and mULC, to support non-standard use cases and applications.

remote assistance. The presented classes of services will be considered due to their profound impact on human lifestyles, as well as their ability to stimulate technological innovation and create new business models, especially as Networking 2030 wireless systems and 6G develop.

4.4.1 Holographic telepresence and conferencing

Holographic telepresence will become one of the main applications of HTC. Ian F. Akyildiz et al. [15] illustrated holographic telepresence as introduced in Figure 4.3. As shown in Figure 4.3, users from different places can meet in the form of holograms. The HTC system at each location includes two main components: a touch booth and a holographic display.

There is another format of holographic telepresence when holograms are generated by computers without the use of sensor booths. As an example,

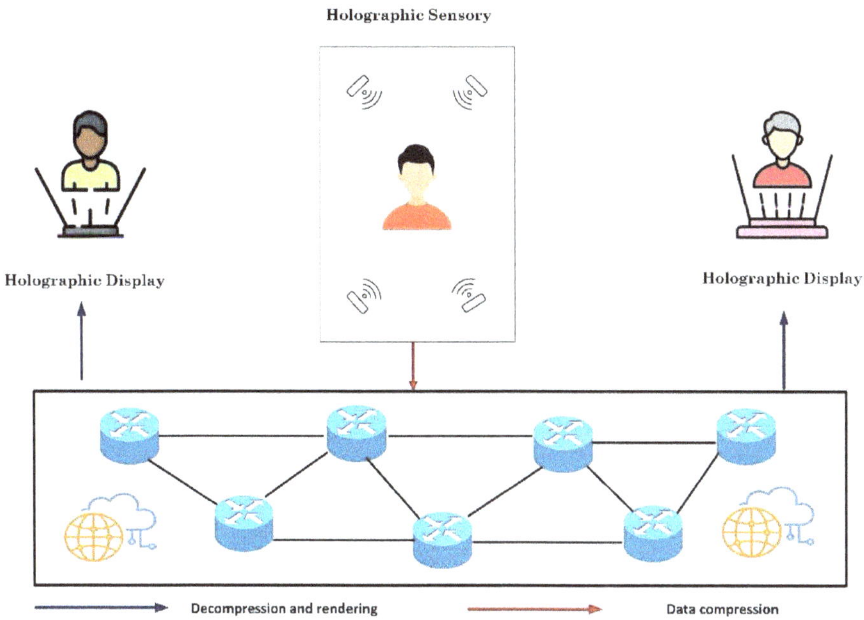

Figure 4.3 Holographic telepresence illustration.

consider Whitney Houston's holographic tour, where three-dimensional objects of the singer are created by computers rather than cameras to record a light field. Several pilot tests of holographic video calls have already been conducted. By leveraging augmented reality, today's video calls could be replaced by holograms that can interact similarly to a physical conversation. Thus, in 2017, the Korean company KT Telecom made a holographic video call between Seoul and New Jersey [24].

Holography has also been used in Spanish projects dealing with "holographic teleportation." Now, the company Newtonlab Space is already testing 5G technology for broadcasting holograms live [25]. A similar project is being developed by Red.es for video conferencing, which will include a combination of augmented reality and holography, as well as a virtual assistant that all participants can see no matter what angle they are standing at.

Online shopping has changed our lives significantly, especially during the pandemic. However, it is always challenging to get an accurate idea about the size, color, smell, and many other attributes of online products just by reading descriptions and viewing images or videos. HTC will allow users to view the product on a real scale, with true colors and 3D geometry [26]. In turn, other multimedia data, such as smells and tastes, can be transmitted. Advanced HTC systems can also allow shoppers to enter a virtual shopping mall with many products. Users wearing smart haptic gloves can pick up items and place them in holographic carts, making this type of virtual shopping hyper-realistic with significant marketing appeal.

Video conferencing has played an important role during the COVID-19 pandemic. It made it possible to unite users worldwide, replacing personal meetings with virtual ones. The format of video conferencing is fundamentally different from real face-to-face conferences. Video conferencing users must look at 2D screens that display remote users at a reduced physical size. Compared to video conferencing, HTC provides a more immersive user experience. The holographic display is large enough to display remote interlocutors at their actual physical size. If the QoE is high enough, participants may not notice the difference between holographic conferences and real in-person ones. With smart gloves, holographic meetings can even support handshakes. An example of a holographic conference is shown in Figure 4.4, where three-dimensional speakers are projected onto a lecture screen in an auditorium at Imperial College London [27].

Holographic technology will also be useful for organizing industrial processes. Through the use of holograms in factories, it will be possible to speed up and simplify the processes of modeling, debugging, manufacturing, and

Figure 4.4 Holographic lecture at Imperial College London (speakers are projected at life size). Taken from [28].

layout of parts. The introduction of holograms into vehicles can bring many new opportunities for users. Envisics has already installed this technology in Jaguar vehicles [29] and hopes to add features such as 3D GPS addresses and pedestrian identification.

4.4.2 Education and training

By analogy with a holographic conference, students and teachers also have the opportunity to organize meetings using HTC technology. Apart from telepresence, educational materials can be converted into holograms to engage students. Nowadays, students need to use their imagination to imagine complex systems, especially 3D concepts such as electromagnetic field propagation or gradients of higher-order functions. With the help of HTC, students can directly observe phenomena in three-dimensional projection. As shown in Figure 4.5, holograms train medical workers using Microsoft HoloLens [30]. Expensive equipment and devices can also be converted into holograms, allowing students to use them virtually without the expense of purchasing them.

4.4.3 Holographic healthcare

Healthcare facilities use holographic projection systems for real-time 3D visualization. The use of holographic tools in the healthcare sector can become a means of providing a visual display of the patient's organs and body, which will allow a specialist to quickly assess his condition in a three-dimensional projection with the possibility of software analysis and spatial zoom [32].

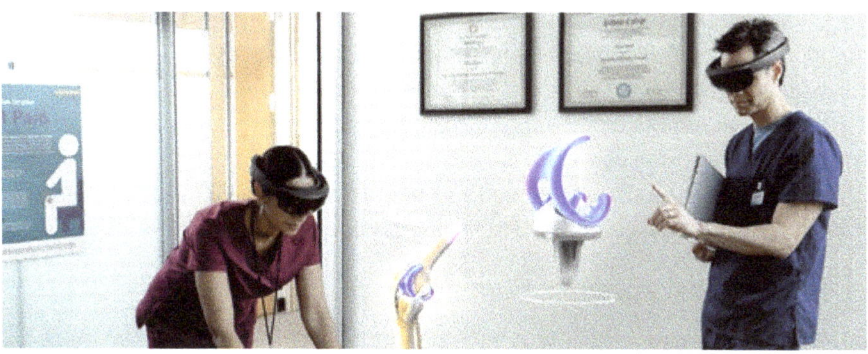

Figure 4.5 Holographic training program developed by Person and Microsoft and used to train healthcare professionals. Taken from [31].

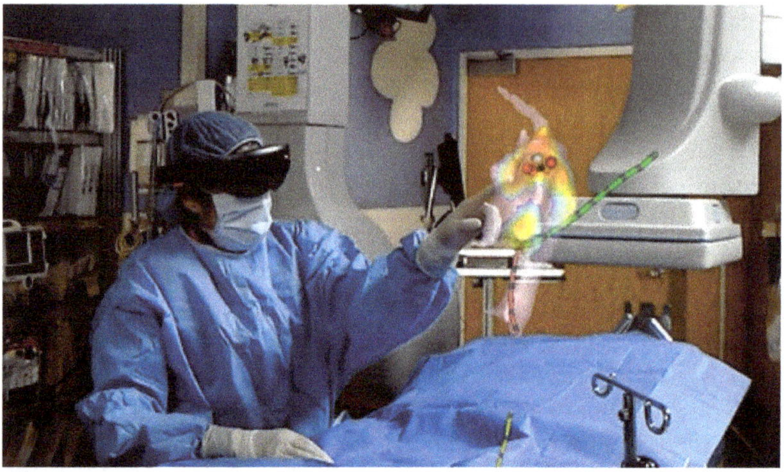

Figure 4.6 A doctor uses a holographic display during a cardiac ablation procedure. Taken from [33].

In other words, HTC can be used for remote interaction with patients. For example, in the case of infectious diseases, the doctor and the examinee can communicate using HTC without direct contact. HTC can provide more complete information than remote monitoring using conventional video services. In addition to observing the patient's face, the doctor can observe his behavior through the holographic display. The doctor can also check the patient's body using smart gloves that provide tactile signals. On the patient side, collecting kinesthetic signals requires arrays of sensors and actuators. Moreover, remote surgery can be performed using robots that copy the actions of a doctor performing surgical procedures directly on a hologram of the patient provided to him. Currently, holograms assist doctors during operations, as shown in Figure 4.6. Remote transmission of holographic images will make the healthcare field more accessible and enhance its effectiveness.

Spain has taken note of the rapid development of this technology and has made significant progress in various sectors. For example, the Sant Pau Hospital in Barcelona has already performed operations using 3D models [34], which allowed surgeons to observe the patient's body in detail during surgery.

4.4.4 Holographic entertainment industry

Thanks to HTC, entertainment will become much more immersive (the effect of immersion in an artificially created environment) than when using 2D screens. Television, sports, and gaming may change radically. Holographic television: holographic content such as news, advertising, and films can be broadcast by providers using HTC technologies. Advanced holographic screens, such as light-field displays, can be used by end users in place of a television and provide a hyper-real 3D stream.

Currently, sports broadcasting is widely used on 2D screens. However, users' focus is limited to the camera's field of view, and there is no way to see other players outside of it. Holographic sports broadcasting can provide a 360° view of the 3D from any angle. For example, a football field can be projected onto a coffee table with a flat holographic display on top [35]. Thanks to this, viewers can see three-dimensional players on a reduced scale across the entire field simultaneously, as if they were in a real stadium. Figure 4.7 shows an example of using XR HMD to display volumetric projections of football players on the table. Similarly, any sports, such as badminton, tennis, and boxing, as shown in Figure 4.8, can be broadcast using holographic communication technology.

Holographic games can be non-immersive or immersive. Non-immersive games use holographic displays, such as holographic flat screens, and require users to use controllers, keyboards, and other tools to provide input. In turn, holograms in non-immersive applications can be perceived by the naked eye or as a result of using XR HMD, in particular, augmented reality (AR) and

Figure 4.7 3D reconstruction of a football match using XR HMDs. Screen taken from the video in [36] at 1.37.

Figure 4.8 Boxing holographic broadcast developed by Condense Reality. Taken from [37].

mixed reality (MR) glasses, as shown in Figure 4.9. Holographic games provide immersion into an artificially created reality due to closed holographic displays that transport users into the virtual world and exclude the visibility of the real environment. It is worth noting that immersive holographic games can also be implemented using VR devices. Users with a VR HMD can view virtual holographic characters and the environment through a near-eye display. A VR HMD differs from a holographic display in that it requires a head-mounted display to receive holographic content, while light-field displays allow anyone near it to perceive holographic images.

4.4.5 Holographic remote support

The HTC remote assistant can find many applications in various areas of human life. So, if it is necessary to provide the user with assistance in repairing equipment, a car, or any devices, remote technical support can only describe the process of solving the problem over the phone. However, difficulties often arise when relating the problem description to its real root

Figure 4.9 Axiom Holographic (formerly Euclideon Holographic), holographic arcade game. Taken from [38].

cause and location. Although video can provide a good demonstration effect, the size and viewing angle of the frame make it difficult to identify and subsequently interpret the problem. HTC, in turn, allows you to display the problem area and how to fix it in a three-dimensional projection with a real scale, allowing users to avoid difficulties in following technical support instructions and increase the efficiency of after-sales customer service.

4.5 Summary

The developing telecommunications industry is creating new types of services that influence the work of enterprises, industries, and everyday processes, and also change people's ideas about the world. These changes are driven by the emergence of new solutions and concepts that affect not only communication networks but also the everyday reality of people. Holographic communication is a technology that is actively developing and contributes to the transition to a new level of knowledge and life, making it more convenient and richer.

In this chapter, a study of the features of the operation of holographic technology and holographic communications was carried out, technologies and devices used to construct and reproduce holographic images were identified, prospects for the development of holographic communication technology were outlined, and the main areas and applications were studied in which the use of holograms can bring positive results and experience.

It was found that to achieve a truly immersive experience, HTC's service needed to utilize all possible human senses. To improve the user experience, in addition to simulating the effects on the main sensory organs, you can also engage the senses of balance and location, as well as provide support for skin sensations such as temperature, wind, humidity, etc. Bringing holograms to the average user will be a major challenge in the future, but several factors are slowing the development of holographic communications today.

One of the main reasons is the complexity and high cost of producing devices for reproducing holograms. Currently, such devices are still considered rare on the market, and their price remains very high. In addition to this, equipment for creating digitized 3D images requires specific knowledge and skills to set up and use correctly. This creates additional difficulties for potential users, especially those who do not have professional information technology skills. In this regard, the development of new technologies and methods to simplify and reduce the cost of hologram playback devices is an urgent task for the telecommunications industry.

In addition, another problem is related to limitations in data transmission technology since high detail requires significant bandwidth. For example, light-field displays with microlens arrays typically require about 10×10 (100 times) more pixels compared to conventional binocular stereo systems. Applications such as 3D reconstruction and telepresence require gigabytes of dense model data, requiring several hundred Gbit/s of bandwidth, which is significantly higher than the bandwidth requirements for video streaming (10–100 Mbit/s) or for augmented and virtual realities (from 100 Mbit to 1 Gbit with a latency of about 5 ms). Insufficient network bandwidth can lead to delays and loss of image quality, which degrades the experience of using holographic communications.

Finally, another factor that slows down HTC's development is the lack of standardization in this area. At the moment, no uniform standards are defining the formats and protocols for transmitting holographic images, which also complicates the development of universal devices and software. For the successful development of holographic communication technologies, a necessary condition is the introduction of completely new ultra-high-density networks that have significant bandwidth and are capable of meeting requirements with ultra-low latency values. With the development of high-performance communication standards, holographic calls may become a reality and become widespread in the near future.

An analysis of existing research has shown that holographic communication has great potential for creating more intuitive, natural, and emotional forms of communication and interaction between people. It can lead to more efficient and productive work in various fields such as education, medicine, science, entertainment, etc. HTC's development also has the potential to create new markets and business opportunities in telecommunications, as it is a promising and rapidly growing sector. All this makes the study and development of holographic communication an important and relevant area in scientific and technological research.

References

[1] A. Clemm, M. T. Vega, H. K. Ravuri, T. Wauters, and F. D. Turck, "Toward Truly Immersive Holographic-Type Communication: Challenges and Solutions," *IEEE Commun. Mag.*, vol. 58, no. 1, pp. 93–99, 2020.

[2] J. Matthews and A. Nairn, *Holographic ABBA: Examining Fan Responses to ABBA's Virtual "Live" Concert*, vol. 46. Popular Music and Society, 2023.

[3] S. Ibuki, T. Okamoto, T. Fujihashi, T. Koike-Akino, and T. Watanabe, "Rateless deep graph joint source channel coding for holographic-type communication," in *GLOBECOM 2023 - 2023 IEEE Global Communications Conference*, 2023, pp. 3330–3335.

[4] L.-H. Lee *et al.*, "All one needs to know about metaverse: A complete survey on technological singularity, virtual ecosystem, and research agenda," *arXiv [cs.CY]*, 2021.

[5] A. Clemm *et al.*, "Rateless deep graph joint source channel coding for holographic-type communication," in *GLOBECOM 2023 - 2023 IEEE Global Communications Conference*, vol. 58, Popular Music and Society, 2020, pp. 3330–3335.

[6] I. Selinis, N. Wang, B. Da, D. Yu, and R. Tafazolli, "On the Internet-scale streaming of holographic-type content with assured user quality of experiences," in *2020 IFIP networking conference (networking)*, IEEE, 2020, pp. 136–144.

[7] "An Evening With Whitney: The Whitney Houston Hologram Concert," *Whitney Houston*. [Online]. Available: https://www.whitneyhouston.com/de/tour/an-evening-with-whitney-the-whitney-houston-hologram-tour/. [Accessed: 20-Apr-2024].

[8] "ABBA Voyage." [Online]. Available: https://abbavoyage.com/. [Accessed: 20-Apr-2024].

[9] I. Fg-Net2030, "Representative use cases and key network requirements for network 2030," FG-NET2030 document NET2030-O-027., 2020.

[10] C. Huang *et al.*, "Holographic MIMO surfaces for 6G wireless networks: Opportunities, challenges, and trends," *IEEE Wirel. Commun.*, vol. 27, no. 5, pp. 118–125, 2020.

[11] A. Covaci, L. Zou, I. Tal, G. M. Muntean, and G. Ghinea, "Is multimedia multisensorial?-a review of mulsemedia systems," *ACM Computing Surveys (CSUR)*, vol. 51, no. 5, pp. 1–35, 2018.

[12] D. Blinder *et al.*, "Signal processing challenges for digital holographic video display systems," *Signal Process. Image Commun.*, vol. 70, pp. 114–130, 2019.

[13] P. A. Kara, A. Cserkaszky, M. G. Martini, A. Barsi, L. Bokor, and T. Balogh, "Evaluation of the concept of dynamic adaptive streaming of light field video," *IEEE Trans. On Broadcast.*, vol. 64, no. 2, pp. 407–421, 2018.

[14] G. Wu *et al.*, "Light field image processing: An overview," *IEEE J. Sel. Top. Signal Process.*, vol. 11, no. 7, pp. 926–954, 2017.

[15] I. F. Akyildiz and H. Guo, "Holographic-type communication: A new challenge for the next decade," *ITU Journal on Future and Evolving Technologies*, vol. 3, no. 2, pp. 421–442, 2022.

[16] T. Bi, A. Pichon, L. Zou, S. Chen, G. Ghinea, and G.-M. Muntean, "A DASH-based mulsemedia adaptive delivery solution," in *Proceedings of the 10th International Workshop on Immersive Mixed and Virtual Environment Systems*, 2018.

[17] "When will holograms enter everyday life?," *VC.RU*. [Online]. Available: https://vc.ru/future/26828-kogda-gologrammy-voydut-v-povsednevnuyu-zhizn. [Accessed: 20-Apr-2024].

[18] S. Schwarzmann, P. Breitbach, T. Zinner, and M. Rost, "Modeling adaptive video streaming using discrete-time analysis," in *2019 31st International Teletraffic Congress (ITC 31)*, 2019.

[19] P. Dhar, "6G's Haptic, Holographic Future? Possibilities and challenges for future 6G communications networks," *IEEEE Spectrum*, 2021. [Online]. Available: https://spectrum.ieee.org/6g-haptic-holography. [Accessed: 20-Apr-2024].

[20] "The holographic and tactile future of 6G. Opportunities and challenges of next generation communication networks," *VC.ru*, 2021. [Online]. Available: https://vc.ru/tech/279402-holograficheskoe-i-taktilnoe-budushchee-6g-vozmozhnosti-i-problemy-setey-svyazi-sleduyushchih-pokoleniy. [Accessed: 20-Apr-2024].

[21] Y. Xu, G. Gui, H. Gacanin, and F. Adachi, "A survey on resource allocation for 5G heterogeneous networks: Current research, future trends, and challenges," *IEEE Commun. Surv. Tutor.*, vol. 23, no. 2, pp. 668–695, 2021.

[22] M. Giordani, M. Polese, M. Mezzavilla, S. Rangan, and M. Zorzi, "Toward 6G Networks: Use Cases and Technologies," *IEEE Commun. Mag.*, vol. 58, no. 3, pp. 55–61, 2020.

[23] M. Z. Chowdhury, M. Shahjalal, S. Ahmed, and Y. M. Jang, "6G wireless communication systems: Applications, requirements, technologies, challenges, and research directions," *IEEE Open J. Commun. Soc.*, vol. 1, pp. 957–975, 2020.

[24] "KT and Verizon tested holographic communication over 5G," *nag.ru*. [Online]. Available: https://nag.ru/news/29709. [Accessed: 20-Apr-2024].

[25] "Newtonlab Space." [Online]. Available: https://en.newtonlabspace.com/. [Accessed: 20-Apr-2024].

[26] "Protohologram." [Online]. Available: https://protohologram.com/ .[Accessed: 15-Apr-2024].

[27] "In London they will start giving lectures using holograms," *tass.ru*. [Online]. Available: https://tass.ru/obschestvo/5753081. [Accessed: 15-Apr-2024].

[28] A. Mckie, "Holograms on campus: good for teaching, good for the environment?," *Times Higher Education*, 2019. [Online]. Available: https://www.timeshighereducation.com/cn/news/holograms-campus-good-teaching-good-environment. [Accessed: 15-Apr-2024].

[29] "A British startup is developing AR displays for cars. General Motors and Hyundai have already invested in it," *RB.RU*, 2020. [Online]. Available: https://rb.ru/story/envisics/. [Accessed: 20-Apr-2024].

[30] "Microsoft HoloLens 2," *Microsoft*. [Online]. Available: https://www.microsoft.com/en/hololens?rtc=1. [Accessed: 10-Apr-2024].

[31] "HoloLens in surgery," *Microsoft*, 2018. [Online]. Available: https://www.microsoft.com/en-gb/industry/blog/health/2018/12/20/ai-healthcare-hololens-surgery/. [Accessed: 20-Apr-2024].

[32] "Bionic eye and 3D bones: medicine of the future, available now." [Online]. Available: https://trends.rbc.ru/trends/industry/6110d2009a79471127d1fc40. [Accessed: 15-Apr-2024].

[33] "WashU-developed holograms help physicians during cardiac procedure," *Washington University in St. Louis*. [Online]. Available: https://source.wustl.edu/2020/07/washu-developed-holograms-help-physicians-during-cardiac-procedure/. [Accessed: 20-Apr-2024].

[34] "World's first surgery using augmented reality (VR)," 2020. [Online]. Available: https://www.meditaly.ru/news/pervaya-v-mire-operacziya-s-ispolzovaniem-dopolnennoj-realnostyu-vr/. [Accessed: 20-Apr-2024].

[35] K. Rematas, I. Kemelmacher-Shlizerman, B. Curless, and S. Seitz, "Soccer on your tabletop," in *2018 IEEE/CVF Conference on Computer Vision and Pattern Recognition*, 2018.

[36] A. Boyle, "Dive into World Cup action: AI researchers turn soccer videos into 3-D 'holograms,'" *GeekWire*, 2018. [Online]. Available: https://www.geekwire.com/2018/dive-world-cup-action-ai-researchers-turn-soccer-videos-3-d-holograms/. [Accessed: 15-Apr-2024].

[37] Oleksandr, "Apple is Going Phygital: Embracing the Future with Vision Pro," 2023. [Online]. Available: https://collectid.io/apple-is-going-phygital-embracing-the-future-with-vision-pro/. [Accessed: 15-Apr-2024].

[38] "Euclideon hologram solution turns the tables on arcade gaming," *blooloop*, 2018. [Online]. Available: https://blooloop.com/play/news/euclideon-hologram-arcade-table/. [Accessed: 10-Apr-2024].

5

Communication Networks for Telepresence and Holographic Communications

5.1 Basic Scenarios for the Exchange of Holographic Information

Holographic content can be pre-recorded and stored on storage devices (HDDs, SSDs, flash memory, etc.), giving users the ability to play 3D images anywhere without requiring end-to-end network segments. However, below, we will consider three main network scenarios for the exchange of holographic information between a data source and one or more recipients.

- Low-interactive HTC can be used for holographic distance learning, ensuring the transfer of educational materials from remote servers. Just like a presentation that needs to change presentation slides every few minutes, holographic content needs to be updated at a low frequency. As a result, an HTC with a low level of interactivity can operate at relatively high latency values.
- Highly interactive HTC is used for online conferencing and remote control, namely in cases that require ultra-low latency, high bandwidth, and high reliability.
- HTC broadcasting: This scenario is used for live broadcasts of sports or public events. At the same time, users receive holographic content without the ability to interact with it, which, of course, expands the boundaries of acceptable latency values. In addition, the dynamically adaptive transmission of information packets can also be used to improve the efficiency of the use of network resources.

5.2 Communication Network for HTC

The general architecture of the HTC holographic communication system has been introduced by F. Akyildiz et al., in [1], as presented in Figure 5.1. It consists of a traffic source, a destination, and a network segment [1].

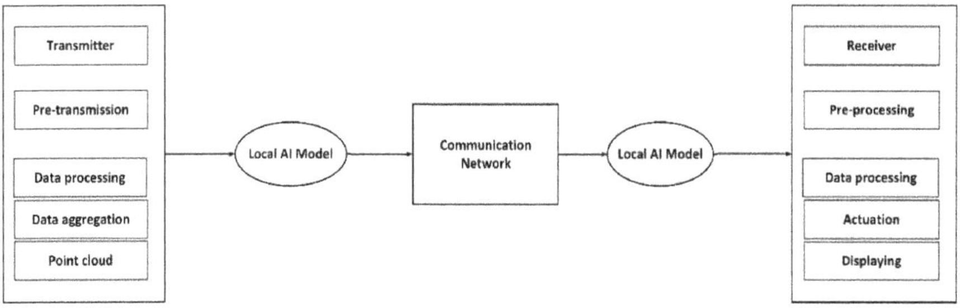

Figure 5.1 HTC system architecture with source and destination computing.

The general architecture of the HTC holographic communication system is shown in Figure 5.1 and consists of a traffic source, a destination, and a network segment [1]. Computing tasks, including modulation, demodulation, encoding, decoding, and synchronization of aggregated data, are mainly performed at the source and destination nodes connected by a telecommunications network.

A. Source

The source mainly performs the following three functions:

1) Uses various sensors to capture images, sound, tactile signals and even smell and taste information.
2) Processes aggregated data. For example, it can create point clouds using images taken from different angles. In addition, the source can synchronize multi-sensor data to ensure correct timestamping. More importantly, the source must support the ability to encode and compress holographic data and media to reduce the amount of network traffic presented to be sent.
3) Uses HTC network protocols to send data packets.

B. Network

HTC networks (wireless or wired) deliver large amounts of source data with guaranteed and limited end-to-end latency. Existing technologies, such as software-defined networking (SDN), automatic network slicing (the ability to use the same physical network to implement many different services), data caching, use of the terahertz range, etc., need to be adapted to meet the requirements of holographic communication [2]. New technologies, such as semantic communication (SC) [3] and deterministic networking (DetNet) [4], modified transport layer solutions, scheduling algorithms to control end-to-end delay, etc., can make HTC networks more reliable and productive.

C. Recipient

The destination receives and then displays the data provided to it. The recipient also has various actuators (olfactory, gustatory, and tactile actuators) to recreate an environment similar to the traffic source. The holographic display, actuators, and speakers are located in different places, but they also need to be synchronized. Synchronization can be carried out directly at the recipient segment or in the communication network, which ensures that data packets are delivered at the scheduled time. In addition to ingesting and rendering content, the recipient can actively correct errors, enhance the quality of holograms and multimedia using ultra-high-resolution technologies, and perform motion recognition for adaptive service requests.

5.2.1 Computing in the network for HTC

The system architecture shown in Figure 5.2 requires the use of large computing resources both at the source and at the recipient. High-performance computers and edge servers can be used to meet service requirements. However, in order to reduce the costs required for the deployment of holographic communication, it is necessary to use inexpensive devices, as well as organize the transfer of calculations to the network using "Computing in the Network" (COIN) technologies [5], as shown in Figure 5.2.

For example, source sensors can communicate directly with actuators, displays, and speakers at the point where traffic is received. Depending on

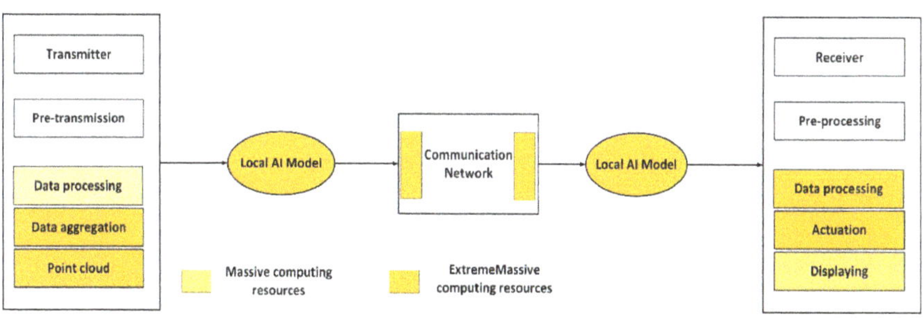

Figure 5.2 Architecture of the HTC system with network computing.

the network condition, the original data may be transmitted without compression, however, factors such as congestion and limited bandwidth require the system to perform adaptive compression directly on the network segment. In addition, the network is capable of unpacking and then sending the visualized data to its destination. Similarly, you can use edge services, cloud servers, and routers with increased performance. Thus, the computational load on the source of holographic traffic and its recipient will be significantly reduced.

5.2.2 HTC broadcasting systems

In addition to end-to-end (one-way) communication, HTC can also support the broadcast of pre-recorded content or "Live" modes such as holographic television, streaming holographic video, and asynchronous holographic education [6]. As shown in Figure 5.3, holographic data is stored on a server or data center, and the NTS performs the function of connecting servers and end users. It is worth noting that service consumers can use a variety of devices, including high-quality light-field displays, XR HMDs, and traditional 2D displays. To meet the needs of a large number of users, holographic content must be individually modified based on QoE requirements.

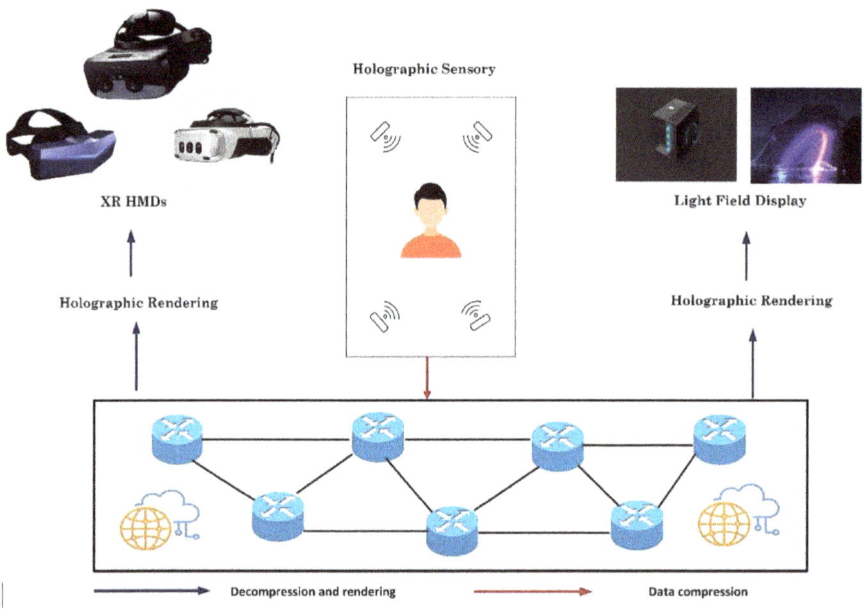

Figure 5.3 HTC broadcasting system with heterogeneous recipients.

5.3 Technical Requirements of HTC

HTC is a complex system that uses a combination of information from many technical areas, such as the physics of displays, sensors, and actuators, as well as data compression methods, wireless communications, computer networks, etc. To develop technical requirements for data networks that support HTC services, it is necessary to analyze existing photo/video formats, introduced in Table 5.1, as well as their parameters [7].

From the data presented in the table, one can note the preservation of the characteristics of previous photo/video formats, simultaneously with the emergence of additional criteria, which indicates a directly proportional increase in the load on the data transmission network in the future.

Next, the requirements and problems associated with HTC coding, communications, and networking will be discussed, which mainly consist of the following four aspects:

- **Data transfer rate:** As discussed earlier, a holographic display requires data transfer rates of up to several terabytes per second. Although efficient coding and COIN distributed network computing can reduce the required data transfer rates to gigabytes per second or even several hundred megabytes per second, the time spent on calculations has a direct impact on increasing network latency. Thus, HTC requires unprecedentedly high data rates, and today's network technologies must be revised and improved to support it. Table 5.2 presents the predicted HTC data

Table 5.1 Parameters of photo, video, 3D video, and hologram formats.

Multimedia item	Formats
Photo	• Image resolution • Color
Video	• Image resolution • Color • Frame rate
3D video	• Image resolution • Image depth • Color • Frame rate
Hologram	• Image resolution • Image depth • Color • Frame rate • The number of viewing angles

Table 5.2 Projected development of holographic communication. Taken from [1].

	Current time	5 years	10 years (6G and above)
Data transfer rate	1 Gbit/s	100 Gbit/s	>1 Tbit/s
Number of frames per second	≤30 fps	60 fps	≥90 fps
Communication range	Local area network (LAN)	Metropolitan area network (MAN)	Wide area network (WAN)

rates, number of frames per second, and communication range. 6G and beyond systems are expected to support high-quality HTC over very large areas [8].

To determine the amount of data transmitted by the camera sensor, Microsoft Kinect [9], which is used further in the study and is capable of creating three-dimensional images, we will carry out some calculations. Each frame has a resolution of 1920 × 1080 pixels with 4 bytes of color information on each of them and also has depth data in Point Cloud format with a resolution of 512 × 424 pixels with 2 bytes of object distance data on each of them. This results in 70.4 Mbits of data per image. Therefore, if such video content needs to be transmitted at 30 frames per second, the minimum network bandwidth must be 2.06 Gbit/s.

It should be noted that to create a full-fledged holographic image, several recording devices are used, which will lead to an increase in network bandwidth requirements in proportion to the number of camera sensors, image resolution, and frame rate. For example, LFV (light-field video) content, which allows for the creation of highly detailed volumetric images, will require bandwidth in the range of 100 Gbit/s to 2 Tbit/s [10].

- **End-to-end delay:** To understand the acceptable level of delivery delay when using HTC services, it is necessary to determine the method of transmitting video material – all the content at once or only the part that the viewer is currently watching from a certain angle. If you choose the first option, the latency requirements will be close to the values necessary for augmented reality devices, but the load on network bandwidth will increase. The second option, on the contrary, reduces the load on the network but increases the latency requirements.

  Delay, in HTC networks, occurs due to the mutual influence of the time required for data collection, encoding, network transmission, decoding,

Table 5.3 Comparison of technical requirements for video, VR, AR&MR, HTC.Taken from [1].

	Video		**VR**	**MR & AR**	**HTH**
Data transfer rate	Resolution	Speed (Mbit/s)	25 (Mbit/s) – 5 (Gbit/s)	25 (Mbit/s) – 5 (Gbit/s)	100 (Gbit/s) – 10 (Tbit/s)
	HD (1080p)	8–12			
	360° (4K)	35–65			
	360° (16K)	100–500			
Latency	15–35 ms		<20 ms	<10 ms	<1 ms
Synchronization	Audio/video		Multiple perspectives	Multiple perspectives	Thousands of perspectives and many sensations
Devices	Smart devices/computers		Head-mounted displays	Head-mounted displays (smart glasses)	Naked-eye or head-mounted displays

display, and actuation of execution mechanisms. For highly interactive applications such as holographic games, the total end-to-end latency should not exceed 20 ms to avoid crashes. Data acquisition, encoding, decoding, and display can take up to 20 ms. As a result, less than 1 ms of the allocated latency resource is left for transmitting traffic over the network. This indicator limits the permissible distance between the source and the recipient since the transmission of information packets cannot exceed the speed of light.

- **Lightweight calculations:** HTC requires a significant amount of computation to optimize transmission. Lightweight computing algorithms and frameworks must be developed to support portable devices that users have ubiquitous access to. For example, a user will be able to project holographic content onto a light field display using a smartphone or tablet. In addition, simplifying computation will reduce end-to-end latency and ensure high quality of service.
- **Multimedia synchronization:** HTC systems have a large number of sensors, actuators, displays, and speakers. Different types of data must be synchronized and presented at the receiving end to avoid any inconsistency. HTC tools must synchronize thousands of different viewing perspectives into a single 3D model, as well as information intended for each of the five main human sensory organs.

Table 5.3 provides a comparison of technical requirements for standard video, VR, AR & MR, and HTC. It is worth noting that in some scenarios, the parameters vary significantly.

5.4 Methods for Obtaining Depth Data on the Traffic Source Side

As discussed earlier, the source basically produces a video consisting of a point cloud. Next, we will look at point cloud capture devices, standards for encoding them, and methods for synchronizing multimedia.

5.4.1 RGB-D devices

The relatively recent advent of affordable consumer RGB-D (red green blue-depth) sensors has unlocked a wealth of functionality for 3D technology. For example, devices such as Microsoft Kinect v1, Microsoft Kinect v2, and Intel RealSense F200 [12] are common RGB-D sensors that use structured light (SL) and time of flight (ToF) technologies to extract depth information for each pixel in the image. Each pixel that is captured by RGB-D devices has two values, color and depth, allowing real-life 3D objects and environments to be digitalized.

Due to the availability of these devices, there has been increased interest in the use of 3D technology in various fields, including healthcare, 3D scanning, modeling, object measurement, robotics, navigation, and others. All three sensors use the projection of the near-infrared (NIR) light at wavelengths of the order of 750–1400 nm to obtain depth data. Kinect v1 (Kinect for Xbox 360) was developed by Microsoft and first introduced at E3 in 2009 under the code name Project Natal [13]. In 2011, the first version of the Kinect SDK was released, and soon after, it began to be used in many 3D research projects. Figure 5.4 presents the front panel of Kinect v1, with the following:

- VGA visible range camera, which is a regular RGB camera with a resolution of 640×480 pixels and a frame rate of 30 frames per second;
- infrared laser projector, which emits structured infrared light to "probe" the space with a grid of dots; and
- infrared camera (CMOS sensor), consisting of a matrix of photodetectors. The monochrome CMOS sensor detects structured light reflected from objects (image of a projected infrared grid).

Kinect v1 uses a 3D scanning technique based on structured light (SL), which involves projecting an infrared pattern of rays (an encoded IR image or grid) using an infrared projector. The rays are then reflected from objects and received by the digital matrix of the CMOS sensor integrated circuit. Brighter areas indicate objects that are close to the camera, and if the object or person is extremely close, the reflection becomes too bright and indistinguishable for the camera.

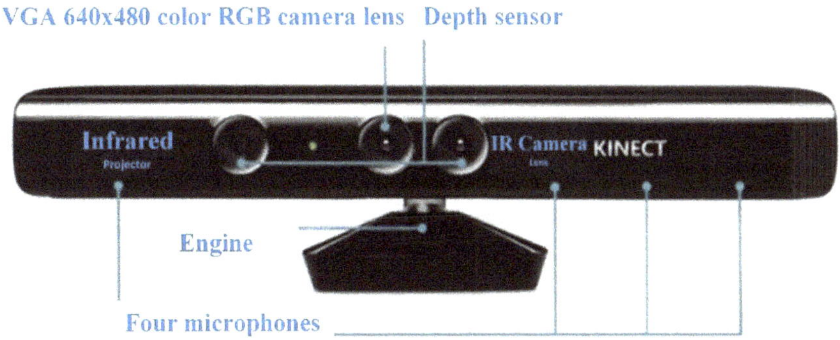

Figure 5.4 Main components of the Microsoft Kinect v1 device.

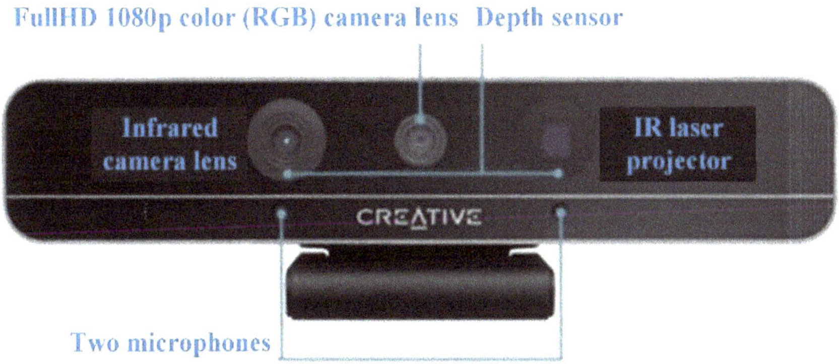

Figure 5.5 Main components of the Intel RealSense F200 device.

The Intel RealSense camera, presented in Figure 5.5, is more compact in size than the Kinect. However, the RealSense F200 is capable of capturing 1080p full HD RGB images as well as 640×480 pixel depth maps. Intel RealSense F200 uses SL technology to create a depth map similar to that used in the first-generation Kinect [14].

As part of the study, the Kinect v2 camera, presented in Figure 5.6, developed by Microsoft and first introduced in 2014 with noticeably higher performance and frame quality compared to Kinect v1, will be used to obtain a point cloud. Instead of the structured light principle used in the first version of the device, Kinect v2 uses time of flight (ToF) technology, which allows it to be used outdoors in bright sunlight [15].

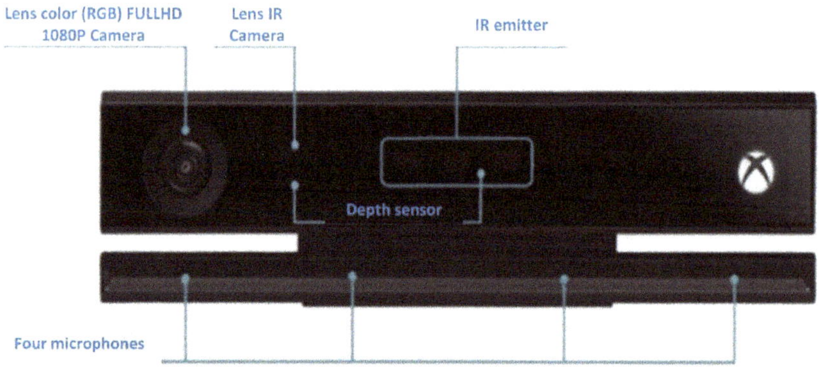

Figure 5.6 Main components of the Microsoft Kinect v2 device.

The ToF technology in Kinect v2 uses constant signal modulation and phase shift of the light wave to obtain depth data, including several stages:

- The infrared emitter projects a continuous modulated wave in the infrared range, invisible to the human eye.
- A continuous wave is reflected from the surface of an object.
- The IR camera lens receives and registers the reflected signal.
- The phase meter measures the phase shift $\Delta\varphi$ of the received signal.
- The distance to the object (depth) is calculated using the following equation:

$$d = \frac{\Delta\varphi}{4\pi f} \cdot c, \tag{5.1}$$

where $\Delta\varphi$ is the phase shift, f is the signal modulation frequency, c is the speed of light (in air it is about 300 thousand km/s), and d is the distance to the object. Figure 5.7 presents the operating principle of ToF technology in Kinect v2.

In other words, the further away an object or component is, the longer it takes for light to reach it, reflect, and return, resulting in an increased phase shift of the received wave.

Figure 5.8 shows several RGB, depth, and IR frames obtained using Kinect v2 (a, b, c) and RealSense F200 (d, e, f) in the same scene. The images show that the quality of RGB-D data produced by Kinect v2 is significantly better than that of the RealSense F200. Kinect v2 is capable of capturing RGB data with higher color accuracy, contrast, and image clarity than RealSense, and the depth data collected with Kinect is denser and contains fewer noisy

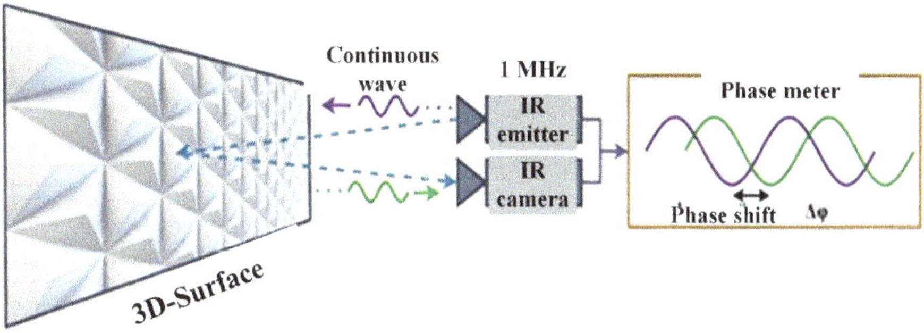

Figure 5.7 The operating principle of ToF technology in Kinect v2.

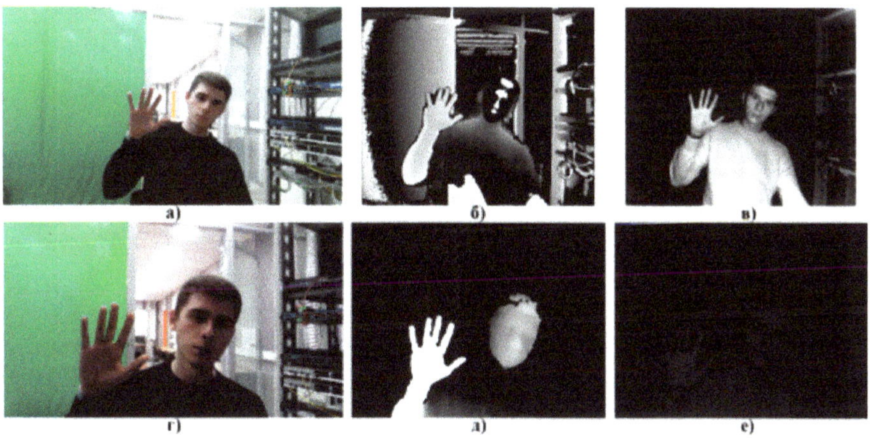

Figure 5.8 RGB, depth, and IR images captured using Kinect v2 and RealSense. F200 on one stage. For Kinect v2: (a) color (RGB); (b) depth map (depth); (c) infrared flux (IR). For RealSense F200: (d) color image; (e) depth map; (e) infrared flux.

pixels, which is a definite plus when working with 3D images. However, Kinect cameras require an external power supply, making them difficult to use on mobile platforms or in power-constrained scenarios. In this case, if you need to use an RGB-D sensor in a portable device or with limited space, the Intel RealSense F200 can meet the requirements due to its small size and light weight.

Table 5.4 contains information about the capabilities and technical characteristics of the RGB-D sensors selected for this study. All information presented has been obtained from operating instructions and manufacturers' official websites.

Table 5.4 Comparative characteristics of RGB-D devices.

Device		Microsoft Kinect v1	Microsoft Kinect v2	Intel RealSense F200
Depth calculation technology		Structured light (SL)	Time of flight (ToF)	Structured light (SL)
Depth camera	Resolution/fps	320 × 240, 30 fps	512 × 424, 30 fps	640 × 480, 30 fps
Color camera (RGB)	Resolution/ fps	640 × 480, 30 fps	1920 × 1080, 30 fps	1920 × 1080, 30 fps
Depth recognition range		0.8–4 m	0.5–4.5 m	0.4–2.8 m
Interface		USB 2.0	USB 3.0	USB 3.0
Dimensions ($W \times H \times D$)		305 × 76.2 × 63.5 mm	249 × 66 × 67 mm	110 × 37.5 × 12.5 mm
Delay		~ 90 ms with processing	~ 60 ms with processing	~ 60 ms with processing
3D scanning of an object		Yes	Yes	Yes
Body tracking		Yes	Yes	No
External power supply		Yes	Yes	No
Power consumption		12 W	15 W	1.7 W

In terms of capturing outdoor depth data, the Kinect v2 is significantly superior to the F200 and Kinect v1. Kinect v2 is also capable of recording depth data on uniformly textured surfaces, something other devices cannot offer. Indoors, the Kinect v2 does a better job of capturing 3D surface data than the Kinect v1 and Intel F200, which similarly represents a significant advantage when working with 3D.

It is important to note that the visual quality of a 3D image is significantly affected by inaccurate depth information, so you should use a device with the best parameters and resolution of the depth sensor and RGB camera. In this regard, Kinect v2 has an advantage over other devices.

5.4.2 Representation and encoding

Several RGB-D depth cameras are organized in an array to collect images of objects from different angles. Based on the collected information and its subsequent processing by specialized software, a three-dimensional point cloud is formed (x, y, z coordinates for each image pixel). To create a three-dimensional image in true colors, it is necessary to combine the depth point cloud represented by the sensor with RGB data received from a color camera.

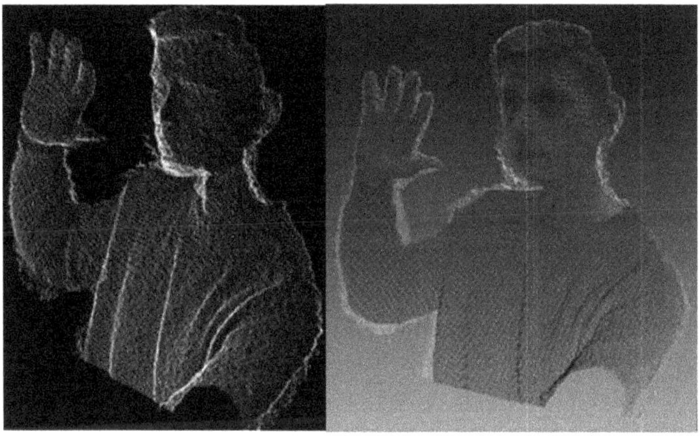

Figure 5.9 A point cloud showing an object in true colors.

The result is a colored object in which each depth value corresponds to a specific color index [16]. Figure 5.9 shows a point cloud showing an object in true colors.

In order to facilitate analysis and reduce network load, a point cloud is often displayed in spectral colors (rainbow colors – close points are depicted as red and yellow gradients and distant points as green and blue) and also requires software compression.

Video-based point cloud compression (MPEG V-PCC) and geometry-based point cloud compression (MPEG G-PCC) are the two main point cloud compression frameworks [17]. V-PCC presents 3D point cloud data as a series of 2D videos, thereby providing full support for the use of existing successful 2D video encoding solutions. G-PCC encodes 3D cloud data directly into 3D space using geometry-based approaches. This method has been used for LiDAR-generated point clouds such as 3D maps. As shown in [18], encoding a point cloud can take several seconds (or even more). Although the size of the compressed data is small, the delay is unacceptable. Existing encoding technologies need to be improved in order to meet the requirements of HTC. In particular, the following three areas of research will solve this problem:

1. **Increasing the compression ratio:** In [19], it was shown that XR video supports significant compression at relatively low quality to reduce network traffic. Super-resolution reconstruction technology based on deep learning is then deployed on the receiver side to improve the quality of the received video. Traditional lossy compression (lossy compression –

the data differs from the original, but the degree of difference is not significant from the point of view of their further use) reduces the quality of the point cloud during subsequent decompression. However, deep learning solutions can overcome this problem by using pre-trained models. In addition, the correlation between adjacent frames can also be used to increase the compression ratio.

2. **Reduced computational complexity during coding:** Increased computational complexity increases end-to-end latency and requires significant computational resources. Therefore, it is necessary to reduce the complexity of encoding calculations. However, there is a relationship between the degree of compression and the amount of computation involved. Typically, a high compression ratio requires high computational complexity, and vice versa. In addition to existing V-PCC solutions, machine learning can be used to train encoding models that will enable efficient extraction of features (individual independent variables that act as inputs to the system). After that, the obtained functions can be sent to the destination, and another machine learning model can be used to restore the original information. Although training may require offline computing resources, real-time online deployment is straightforward.
3. **Use of negotiation between source and recipient**: Information about the user's gaze direction of holographic content is useful for efficient point cloud encoding at the source side. So, for example, more data can be placed in the primary field of view of the recipient user. In other words, by using feedback information to schedule transmitted data, the redundant information required for subsequent encoding and decoding can be greatly reduced.

5.4.3 Media synchronization

Multimedia includes many types of sensors located in different places and requiring mandatory synchronization [20]. However, time coordination is difficult for the following two reasons:

1. Different types of sensors have different response times. For example, cameras can support shooting speeds of around 50 frames/images per second, which results in a capture latency of 20 ms, while microphones can have a latency of several milliseconds, which is much less than that of cameras. For highly dynamic tactile sensors, the delay should not exceed 1 ms [21]. Consequently, various recording devices need to be calibrated and synchronized, considering the complex heterogeneity of

hardware functionality and support. Machine learning algorithms can also be used as a solution. Typically, a given behavior or activity involves several interrelated effects on the senses. Machine learning classifiers (algorithms that are used to map input data to a specific category) and predictors (predictive variables) can be used to check the consistency of collected sensations and automatically generate missing or delayed factors affecting a person's perception.

2. Human perception has different levels of tolerance to delays for certain senses. As the study [20] shows, the airflow sensation delay can be as high as 1 ms to provide an acceptable user experience. However, this certainly depends on the application and the distance between the person and the actuators. For time-sensitive systems such as remote surgery, the latency of haptic signals should not exceed 1 ms.

5.5 Functional Requirements for the Recipient of Holographic Traffic

The receiver plays the HTC content and provides feedback to the source to improve and maintain high-quality QoE. Specifically, the functionality of the receiving segment includes the following:

1) Detecting and correcting errors, improving the resolution and quality of holograms and multimedia, and synchronizing received packages from several sources.
2) Monitoring network status and determining the desired quality of service. Such information can be sent back to the source to optimize encoding. For example, when there is no congestion in the network and there is sufficient bandwidth, the recipient can request high-quality HTC content with the desired QoE. Alternatively, when network capacity is insufficient, the traffic ingestion points request semantic context without other detailed data and can use AI-enabled fine-tuning and holographic enhancement technology. For example, during a telepresence conference, users' gaze is most often focused only on the hologram's face. All other parts of the body can be independently set to a lower resolution.

The problem under consideration also includes predicting the movement of the recipient user using artificial intelligence in order to organize adaptive data transmission by the source and reduce end-to-end delay. Thus, if the source has information about the direction of the recipient's gaze, high-resolution data can be transmitted to a specifically specified field of view

(FoV) of the target user. Motion prediction uses a variety of sensors, such as cameras and inertial measurement devices, that track head, arm, and body movements.

For XR users, motion sensors are integrated into the HMD. However, due to the fact that light-field displays allow holographic content to be observed with the naked eye, it is difficult to find optimal sensor placements. On the one hand, if the sensors are placed far from the user, the measurement accuracy cannot be guaranteed. On the other hand, if the sensors are wearable, the quality of service will decrease since the user will have to use additional devices.

Next, we will consider the functional requirements for the terminal devices used in the study for reproducing holographic images (3D projector, holographic fan, mixed-reality glasses, and holographic pyramid).

5.5.1 Functional requirements for the 3D projector

Functional requirements for a 3D projector include:

1) Display images in 3D format with support for various standards (RealD, Dolby 3D, etc.).
2) High matrix resolution (1920 × 1080 – full HD and higher).
3) Availability of inputs for connecting to various video sources (HDMI, VGA, USB, etc.).
4) Ability to adjust brightness (1500 ANSI lm), contrast, and focal length to ensure optimal image quality.
5) Network connectivity (Ethernet – RJ45, built-in Wi-Fi 2.4 and 5 GHz).

5.5.2 Functional requirements for a holographic fan

Functional requirements for a holographic fan include:

1) The ability to create a three-dimensional image with a wide viewing angle (160° and more).
2) The image created by a holographic fan must be clear, rich, and contrasting (the number of RGB LEDs on the blades is 720 and higher, and the brightness is 1400 cd/m^2 and higher).
3) Supports two main connection options:
    - Local mode, in which the device acts as an access point, creating its own Wi-Fi network (2.4G 802.11b/g/n).
    - Network mode, in which the device connects to a pre-configured Wi-Fi network.

4) The ability to remotely control image settings, including changing the tilt angle, blade rotation speed, and LED brightness.
5) Support for various image and video formats (image: jpg, jpeg, png, gif; video: mp4, avi, rmvb, mkv, wmv, mov, flv).
6) Low noise level (less than 35−40 dB).
7) Easy to install and use.

5.5.3 Functional requirements for mixed-reality glasses

Functional requirements for mixed-reality glasses include:

1) Sufficiently high resolution (2K 3:2 and higher) of holographic lenses (waveguides) to create a clear image of the virtual environment in combination with the real world.
2) The ability to track head movements with sensors (four cameras sensitive to visible light) and simultaneously support control of the direction of the user's gaze (2 IR cameras) in three dimensions for the precise placement and movement of virtual objects in the real world.
3) Accurate and reliable operation of gesture recognition systems (fully rotatable model with two hands, direct interaction) and objects using built-in time-of-flight depth sensors (resolution not lower than 1 MP).
4) The device is lightweight and comfortable to wear for a long time.
5) Ability to connect to the internet and other devices via Wi-Fi (5 GHz 802.11 AC 2 × 2), Bluetooth 5.0, or USB-C interfaces.
6) Sufficient performance to perform complex computing tasks associated with processing and displaying virtual objects (computing platform no lower than Qualcomm Snapdragon 850).
7) Acceptable battery life of the device (2–3 hours of active use).

5.5.4 Functional requirements for the holographic pyramid

Functional requirements for a holographic pyramid include:

1) Use in the manufacture of high-quality glass or transparent acrylic, providing high refractive power and the absence of image distortion.
2) The dimensions of the holographic pyramid must be adjusted to the diagonal of the projection device (with respect to the proportions and the angle of inclination of the edges of 45°).
3) Make the angle of reflection equal to the angle of incidence, which is in the same plane.

4) Smooth, defect-free surface of the pyramid (since any irregularities or scratches can significantly affect the quality of the hologram display).
5) High assembly accuracy (errors in manufacturing and gluing projection edges can lead to distortion, loss of clarity, and brightness of the image).
6) Use of flat projection monitors (1080p and higher) with high brightness (400 cd/m^2 and higher) or TVs with a resolution of 4K and higher, supporting OLED and QLED technologies.
7) Good structural stability and transportability for use in various conditions.

5.6 Summary

This chapter analyzed the main scenarios for the exchange of holographic information and identified three methods: low-interactive for holographic distance learning, highly interactive for online conferences and remote control, as well as the HTC broadcasting scenario for live broadcasts of public events. Each scenario has its own characteristics and requirements for network parameters, such as latency, bandwidth, and packet delivery reliability.

The HTC holographic communication system includes a source, a recipient, and a network segment. The source captures multisensory data, processes it, encodes it, and sends it using HTC protocols. HTC networks deliver large volumes of data with guaranteed end-to-end latency. The receiver displays the data and has various actuators to recreate the environment. Synchronization can be carried out directly in the recipient segment or in the communication network. The recipient can also enhance the quality of holograms and multimedia, as well as perform motion recognition for adaptive service requests. New technologies such as semantic communication (SC) and deterministic networking (DetNet) can make HTC networks more reliable and productive.

A comparative analysis of RGB-D devices Microsoft Kinect v1/v2 and Intel RealSense F200 showed that the quality of RGB-D data obtained using Kinect v2 is significantly superior to that of RealSense F200 and Kinect v1. Kinect v2 has fewer noisy pixels, denser depth readings, better depth mapping accuracy for outdoor objects, and the ability to capture depth values from uniformly textured surfaces.

HTC is a complex system that uses many technical areas, such as the physics of displays, sensors, actuators, data compression, wireless communications, and computer networks. The HTC system requires high data rates, minimal latency, multimedia synchronization, and optimization of computing algorithms to maintain high quality of service. In order to achieve these requirements, it is necessary to improve existing network technologies and develop new, more productive systems. It is expected that 6G and higher communication networks will be able to provide ubiquitous high-quality HTC with adequate QoS and QoE.

References

[1] I. F. Akyildiz and H. Guo, "Holographic-type communication: A new challenge for the next decade," ITU Journal on Future and Evolving Technologies, vol. 3, no. 2, pp. 421-442, 2022.

[2] A. Ateya *et al.*, "Model mediation to overcome light limitations—toward a secure tactile Internet system," *J. Sens. Actuator Netw.*, vol. 8, no. 1, p. 6, 2019.

[3] Y. Liu, X. Wang, Z. Ning, M. Zhou, L. Guo, and B. Jedari, "A survey on semantic communications: technologies, solutions, applications and challenges," *Digit. Commun. Netw.*, 2023.

[4] B. Liu *et al.*, "Towards large-scale deterministic IP networks," in *2021 IFIP Networking Conference (IFIP Networking)*, 2021.

[5] A. Sapio, I. Abdelaziz, A. Aldilaijan, M. Canini, and P. Kalnis, "In-network computation is a dumb idea whose time has come," in *Proceedings of the 16th ACM Workshop on Hot Topics in Networks*, 2017.

[6] I. F. Akyildiz, A. Kak, and S. Nie, "6G and beyond: The future of wireless communications systems," *IEEE Access*, vol. 8, pp. 133995–134030, 2020.

[7] "How to choose the best video format," *movavi*, 2024. [Online]. Available: https://www.movavi.ru/support/how-to/what-video-format-is-the-best.html. [Accessed: 25-Apr-2024].

[8] D. Xu *et al.*, "Edge intelligence: Empowering intelligence to the edge of network," *Proc. IEEE Inst. Electr. Electron. Eng.*, vol. 109, no. 11, pp. 1778–1837, 2021.

[9] S. G. Heng, R. Samad, M. Mustafa, N. R. H. Abdullah, and D. Pebrianti, "Analysis of performance between Kinect v1 and Kinect v2 for various facial part movements," in *2019 IEEE 9th International Conference on System Engineering and Technology (ICSET)*, 2019.

[10] J. Lv, F. Dai, Q. Zhao, H. Li, Y. Ma, and Y. Zhang, "Panoramic light field video acquisition," in *2018 IEEE International Conference on Multimedia and Expo (ICME)*, 2018.

[11] A. Karakottas, A. Papachristou, A. Doumanoqlou, N. Zioulis, D. Zarpalas, and P. Daras, "Augmented VR," in *2018 IEEE Conference on Virtual Reality and 3D User Interfaces (VR)*, 2018.

[12] Y. W. Kuan, N. O. Ee, and L. S. Wei, "Comparative Study of Intel R200, Kinect v2, and Primesense RGB-D Sensors Performance Outdoors," *IEEE Sens. J.*, vol. 19, no. 19, pp. 8741–8750, 2019.

[13] "Kinect Chronicles [Part 1: Birth of Natal]," *xboxunion*, 2020. [Online]. Available: https://xboxunion.ru/article/hroniki-kinect-chast-1-rozhdenie/. [Accessed: 10-Mar-2024].

[14] P. Henry, M. Krainin, E. Herbst, X. Ren, and D. Fox, "RGB-D mapping: Using Kinect-style depth cameras for dense 3D modeling of indoor environments," *Int. J. Rob. Res.*, vol. 31, no. 5, pp. 647–663, 2012.

[15] 李占利 Li Zhanli, 周 康 Zhou Kang, 牟 琦 Mu Qi, and 李洪安 Li Hong/an, "TOF camera real-time high precision depth error compensation method," *Infrared Laser Eng./Hongwai Yu Jiguang Gongcheng*, vol. 48, no. 12, p. 1213004, 2019.

[16] M. S. A. Manap, R. Sahak, A. Zabidi, I. Yassin, and N. M. Tahir, "Object detection using depth information from Kinect sensor," in *2015 IEEE 11th International Colloquium on Signal Processing & Its Applications (CSPA)*, 2015.

[17] C. Cao, M. Preda, and T. Zaharia, "3D point cloud compression: A survey," in *Proceedings of the 24th International Conference on 3D Web Technology*, 2019, pp. 1–9.

[18] H. Liu, H. Yuan, Q. Liu, J. Hou, and J. Liu, "A comprehensive study and comparison of core technologies for MPEG 3-D point cloud compression," *IEEE Trans. On Broadcast.*, vol. 66, no. 3, pp. 701–717, 2020.

[19] M. Dasari, A. Bhattacharya, S. Vargas, P. Sahu, A. Balasubramanian, and S. R. Das, "Streaming 360-degree videos using super-resolution," in *IEEE INFOCOM 2020 - IEEE Conference on Computer Communications*, 2020.

[20] Z. Yuan, T. Bi, G.-M. Muntean, and G. Ghinea, "Perceived synchronization of mulsemedia services," *IEEE Trans. Multimedia*, vol. 17, no. 7, pp. 957–966, 2015.

[21] K. Antonakoglou, X. Xu, E. Steinbach, T. Mahmoodi, and M. Dohler, "Toward haptic communications over the 5G tactile internet," *IEEE Commun. Surv. Tutor.*, vol. 20, no. 4, pp. 3034–3059, 2018.

6

Development of Test Scenarios and Methods for Transmitting Holographic Images Through a Model Network Segment

To form a three-dimensional image, you must go through several mandatory steps. First, continuous shooting of the user is carried out in 360° using software and an RGB-D camera, or the option of creating a sequence of frames with an offset of 15°–30° and their subsequent gluing, alignment, and debugging is allowed. After that, the created hologram image is transmitted through communication networks to reproducing devices, such as MR glasses (covering augmented and virtual reality), 3D-enabled projectors, holographic fans and truncated pyramids, AR smartphones, etc.

6.1 Network Protocol Stack and Key Technologies

HTC network technologies must support high data rates while maintaining guaranteed and limited end-to-end latency. However, network throughput and end-to-end latency cannot exist independently of each other due to existing limitations of network protocols. Next, the general stacks of HTC network protocols will be presented, and then each of the layers will be discussed in more detail separately. Although HTC is introducing many new technologies that can transform existing computer networks, the holographic communication service still needs to be gradually developed based on existing network protocols. Figure 6.1 presents the key developing technologies that will be used in the widespread implementation of scientific and technological progress.

The HTC use cases discussed earlier are supported by technologies such as semantic communication, XR, multimedia, and point cloud encoding and decoding. Transmission control protocol (TCP) and user datagram protocol (UDP) are widely used transport layer protocols. TCP-based protocols are reliable but have significant end-to-end latency. In turn, in UDP-based

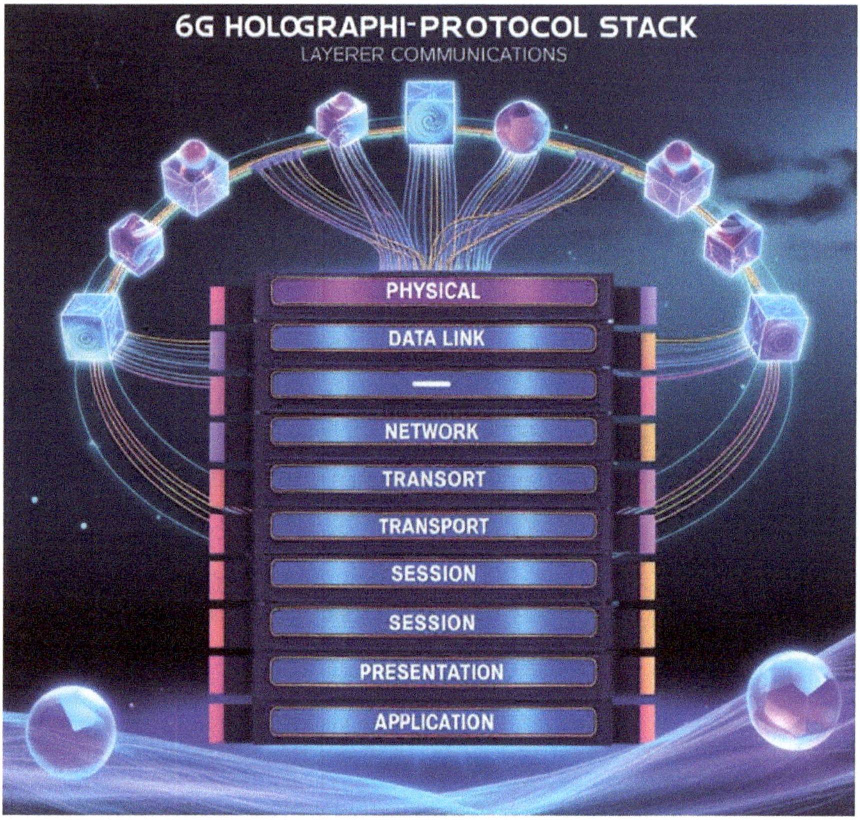

Figure 6.1 Network protocol stack and key technologies for HTC.

protocols, packet loss is possible, but the latency is low. Streaming HTC content poses significant challenges; so it is critical to develop and define optimal and fast transport layer protocols. The network layer also requires the introduction of new technologies, such as New Internet Protocol (IP) (a network tool that provides high-precision service attributes) [1].

Packet Wash is a recently proposed technology that allows selective prioritization of payload fragments within the packets themselves. The main goal is to reduce the number of dropped packets when the network is congested by providing the ability to drop less important parts of the payload first, deterministic networks, and network computing to effectively reduce end-to-end latency and support source encoding and target decoding. The physical and data link layers use 6G and higher technologies, including terahertz communications, visible light communication (VLC), and others. In addition, joint (or coordinated) wireless communications and wireless sensing (a technology in which wireless signals can determine the state of surrounding objects using reflection, diffraction, and scattering when propagating in space) will play an important role in the development of scientific technology. Wireless sensing can eliminate the need for users to wear sensors, allowing for greater flexibility and ease of use. In addition, artificial intelligence can be applied at each level to provide predictive and classification capabilities.

Current telecommunications networks are unable to support high-quality HTC. For example, if TCP is used to stream holographic communication content, then the required bandwidth can be calculated using the following formula [1].

$$T \leq \frac{\mathrm{MSS}}{\mathrm{RTT}} \frac{C}{\sqrt{\rho}}, \tag{6.1}$$

where MSS is the maximum segment size, RTT is the round-trip time, and ρ is the percentage of packet loss. For simplicity, the constant C is implicitly assumed to be equal to one but may be equal to other values in the region of 1, depending on the type of assumption. Existing communication networks cannot universally support such short round-trip times and high delivery reliability. Also, do not forget that there is a physical limitation of RTT due to the constant speed of signal propagation. For example, within 1 ms, at the speed of light, a signal can only travel over a distance of 300 km [2].

The development of HTC network protocols can lead to a new network paradigm. In other words, if successful, the network protocol stacks in Figure 6.1 will gradually turn into the stacks shown in Figure 6.2. Holographic and multimedia technologies support holographic applications. The source and recipient must negotiate and select the quality of the holograms and the number of senses involved that can be supported based on their available hardware. Holographic teleport (holoportation – transmission with near real-time latency – less than 1 ms) [3] can provide various services to meet low latency and high throughput requirements. Teleport is more focused on real-time applications than transport. New IP, in turn, can forward packets from source to destination with a predefined and guaranteed end-to-end delay. The layers discussed are supported by wireless or wired physical communication.

6.2 HTC Services Transport Layer Protocols

TCP made it possible to provide a reliable connection between the source and the destination, i.e., provided a low packet loss rate during the process of loading a 3D object onto a holographic fan. However, TCP has higher

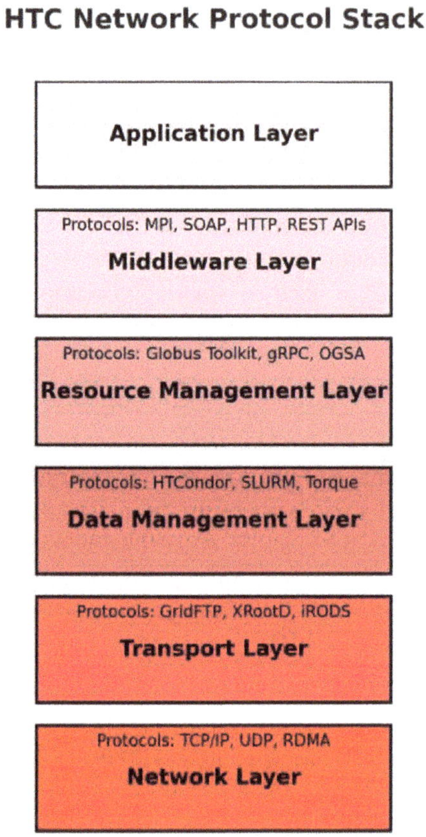

Figure 6.2 HTC network protocol stacks.

end-to-end latency than UDP. UDP does not organize a reliable connection due to the lack of handshaking between the source and the recipient (there is no preliminary establishment of a connection and confirmation of the reception and transmission of a structural unit of information). As a result, the packet loss rate is higher, and the destination must resort to recovering the lost information.

The following are promising streaming technologies and protocols for HTC services:

- MPEG dynamic adaptive streaming over HTTP is an adaptive streaming technology that provides the ability to deliver streaming multimedia content over the internet using the HTTP protocol. MPEG-DASH is a promising standard for low-interactive HTC applications [4]. The technology involves breaking content into a sequence of small file segments, each of which contains a limited encoded portion of the content with varying quality. For point cloud videos, segmentation can be done in space and time. The recipient, in turn, has the opportunity to select (from available alternatives) the optimal segment with the highest bitrate, depending on the assessment of the network condition, without causing freezes and buffering. The protocol is based on TCP and can use existing HTTP web infrastructure to deliver content. However, storing MPEG-DASH data in different qualities for HTS applications on servers requires a huge amount of memory.

- UDP-based protocols, such as Web Real-Time Communications (WebRTC – transfer of streaming data in real time between browsers or other applications that support it using point-to-point technology), need further study and refinement for transmitting HTC data [5]. They have the potential to support highly interactive HTC applications. QUIC (Quick UDP Internet Connections) was used to transmit 360° video in XR systems and allows multiplexing of multiple data streams using priority schedulers and providing low end-to-end latency. Also, it contains encryption capabilities and tolerates packet loss well through edge alignment cryptographic blocks along packet boundaries. Priority transmission is necessary to meet latency requirements, i.e., data packets with high statuses can obtain network resources sufficiently. QUIC will become an important transport protocol in HTC, allowing applications to save network traffic. Holograms can be divided into tiles (from the English tiles – tiles, areas of a certain size), and then only those tiles that have high priority and are in the field of view of the target user can be updated.

Because HTC is an emerging technology, there are currently no well-defined and widely used streaming protocols. TCP-based solutions require the allocation of network resources and network traffic information to reduce end-to-end latency, while UDP-based solutions require the development of packet loss recovery techniques.

6.3 HTC Service Network Layer Protocols

In this work, the IP will be responsible for delivering data from the source to the target system in the model network, but in the future, the network layer of the HTC should be well programmable, reconfigurable, and adaptable, which can be achieved using the New IP concept [6].

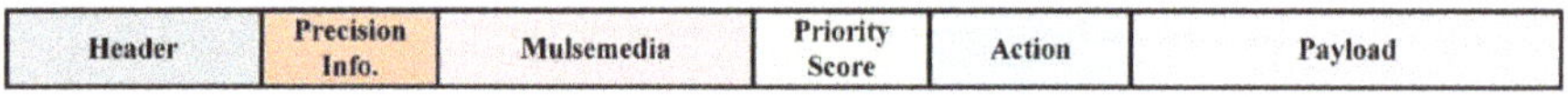

Figure 6.3 Structure of a New IP type packet for HTC networks using semantic communication and packet washing.

New IP allows packets to be delivered based on service requirements that are defined in the packet configuration. In [7], the authors developed a qualitative communication framework, in which packets are sent with varying importance and then selectively discarded based on their congestion priority values. This service allows packet payloads to be broken down into smaller units (called chunks), providing much finer granularity in bandwidth usage. The significance indicator can be included in New IP packets, an example of the structure of which is shown in Figure 6.3.

The packet provides network routers with clear information about how and when it needs to be delivered to its destination. Packets can be discarded depending on their priority value using packet wash when the network is congested [8]. New IP also offers ways to implement semantic communication networks. New IP performance provides reconfigurability and flexibility to deliver large data packets with guaranteed latency requirements.

New IP can be used to deliver DASH or semantic information depending on the priorities set. For example, HTC packets can be classified based on the level of semantic information. In other words, if a frame contains most of the source information, its priority class is assigned a high score, and vice versa. If the packet does not carry meaningful source information but includes some reference data, its priority value is small. When network capacity is sufficient, raw data packets can be sent along with semantic (meaningful) packets, while when there is network congestion, these frames can be gradually discarded as they move toward their destination. Adaptive semantic communication must be designed to take full advantage of network resources to provide optimal QoE. At the destination, dropped packets with low significance can be recovered using deep learning.

In addition to new protocols at the network level, new services are also needed to support HTC applications. End-to-end delay in existing telecommunications networks is a random variable, and fluctuations can be significant. Traditional multimedia communication is capable of buffering data for 100 ms for decoding and rendering to ensure a smooth user experience. However, 100 ms of HTC streaming generates a large amount of data that is often difficult or impossible to buffer. It follows that packets must arrive at their destination within a guaranteed time frame so that they can be processed immediately without consuming buffer space. This requires a deterministic network in which packets can be delivered on a schedule. In [1, 9], three different time-related services are proposed for the 2030 network:

- **Guarantees of early delivery:** Data packets arrive before a certain time, as shown in Figure 6.4.

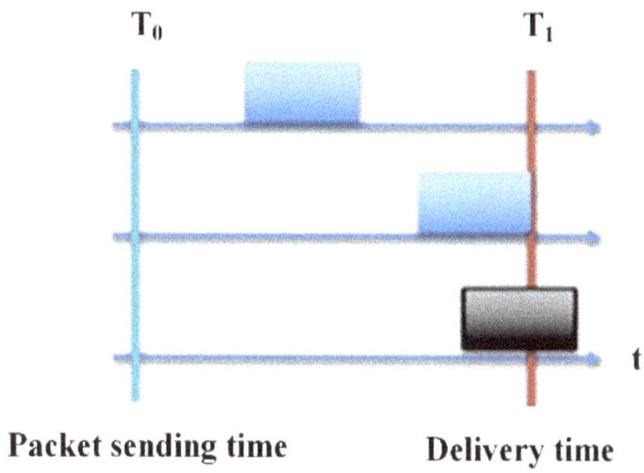

Figure 6.4 The package delivery deadline is T_1. The top and middle packages are delivered ahead of schedule, which meets the requirements. The bottom packet is fully delivered after the specified value. Taken from [1]

- **Guarantees of timely delivery:** Data packets arrive at a certain time with guaranteed deviations, as shown in Figure 6.5.

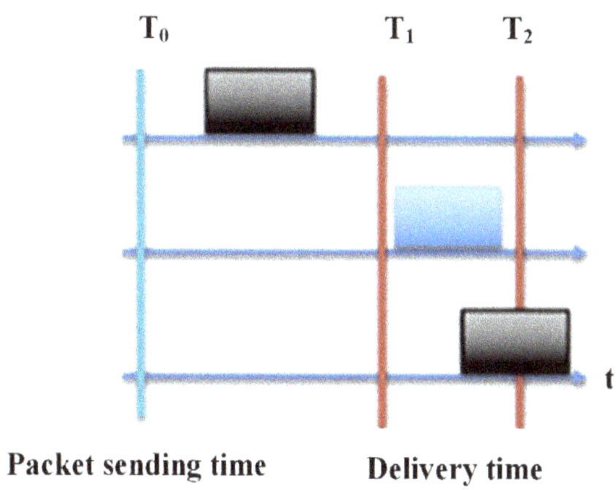

Figure 6.5 The specified packet delivery time is between T_1 and T_2. The top package is delivered earlier, and the bottom package later. The package in the middle is delivered on time, which meets the requirement. Taken from [1]

- **Agreed guarantees:** Multiple streams of information packets arrive with guarantees of early or on-time delivery, as shown in Figure 6.6.

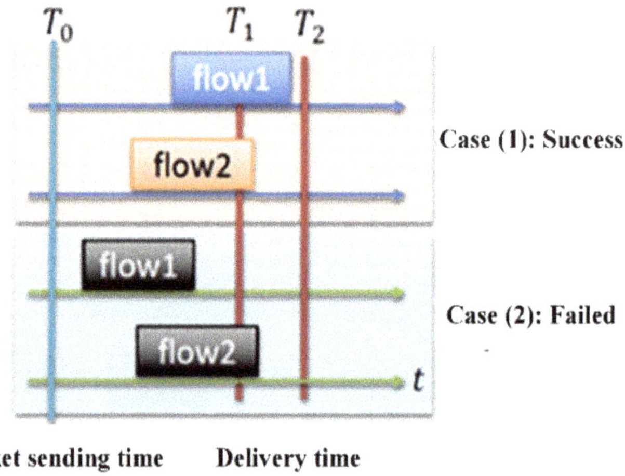

Figure 6.6 The top two streams of packets are guaranteed to be delivered on time, but the bottom two streams are not. Taken from [1]

An on-time guarantee is more challenging than an advance guarantee because packages cannot be delivered before the due date. Existing networks are currently unable to provide the types of services in question. SDN, network function virtualization, and automatic network slicing are key enablers for these services. For HTC networks, agreed delivery guarantees are of paramount importance. Multimedia can include multiple streams of diverse information intended for different senses. They must be delivered on time with the agreed guarantee. In case any stream is delivered early, it needs to be buffered. The volume of holographic and tactile data is large and requires significant storage resources. In contrast, if all streams are delivered on time, the recipient can immediately synthesize and visualize the received multimedia information.

The deterministic network (DetNet) architecture is proposed by the IETF DetNet working group to provide bounded network latency and reduced packet loss for real-time computing applications [10]. This technology can reduce unwanted randomness and delivery variability in networks. Time-sensitive networking (TSN), created by the IEEE 802.1 Standards Working Group, is a technology that aims to provide zero congestion loss and limited latency by reserving network resources and providing network interoperability with strict time synchronization [11]. The development and implementation of DetNet and TSN can facilitate the design of deterministic networks for HTC.

6.4 Physical and Link Layers of HTC Services

The physical and data link layers face challenges in providing high data rates with low latency and high reliability. Terahertz communication is the main technology that will be used in holographic communication services due to its wide bandwidth. It can provide data rates of several terabits per second, which meets the requirements of high-quality light-field displays [12]. In addition to research tasks at the physical and link levels of wireless systems 6G and above, HTC faces the following unique challenges:

- **High energy and spectral efficiency:** Large volumes of hologram and multimedia data require extremely high energy efficiency of communication systems, especially for battery-powered devices. The high energy efficiency of communication systems avoids frequent replacement/recharging of batteries. Terahertz communication allows for high energy efficiency at ultra-high data rates. Recently, THz communication systems have been used for XR applications, where servers and head-mounted displays (HMDs) communicate in the THz range [13]. Subsequent research will be able to improve the energy and spectral efficiency further to achieve data rates above 1 Tbit/s and meet all the requirements of multimedia holographic traffic.
- **High reliability:** Terahertz communication signals are highly directional and prone to blocking. User mobility and a dynamically changing environment can affect the reliability of communications. Reconfigurable intelligent surfaces (RIS) in the terahertz range consist of several, mostly passive elements that can control reflection characteristics and direct (focus) wireless signals in optimal directions, thereby avoiding blockages, leveling out interference, and improving received signals. Although reconfigurable terahertz smart surfaces have proven their effectiveness, it is unclear how to use these systems for HTC applications optimally in different scenarios.
- **Coexistence of wireless communication and wireless sensing:** Wireless sensing (reconstruction of the environment based on the reflection, diffraction, and scattering of wireless signals during propagation) is an integral part of HTC systems since it can replace wearable sensors for

tracking the position of the hands, head, and body. The data link layer must allow the coexistence of wireless communications and wireless sensing. Since the terahertz band has a wide bandwidth, a small sub-band can be allocated for wireless sensing. In addition, THz RIS can function as a sensor hub that passively monitors the environment based on incident and reflected waves [14].

At the current stage, the expected network characteristics, necessary technologies, and traffic parameters for holographic applications have been determined. Next, it is necessary to develop a model for the transmission of holographic copies to verify the compliance of the declared parameters experimentally.

6.5 Laboratory Bench Structure

During the research process, it is necessary to capture a detailed analysis of HTC holographic service traffic to determine threshold parameter values that directly affect the levels of quality of service (QoS) and experience (QoE), as well as the subjective indicator, which, for the client, is of paramount importance.

As part of the testing of "Network technologies with ultra-low latency and ultra-high density based on the widespread use of artificial intelligence for 6G networks," the Department of Communication Networks and Data at St. Petersburg State University of Telecommunications developed a testing framework as presented in Figure 6.7.

The ongoing research involves testing and assessing the quality of perception of holographic images for four devices:

1) 3D projector
2) Holographic fan
3) Mixed-reality glasses
4) Holographic truncated pyramid

Using technical means available in the laboratory, a local connection was established between the RGB-D (red green blue-depth) Kinect v2 camera, the LG CineBeam HF85LSR projector with 3D support, the DseeLab-65X holographic fan, and the HoloLens 2 Development Edition mixed-reality glasses. Figure 6.8 presents the proposed network model for exchanging holographic data.

To create a laboratory stand for the exchange of holographic data, dense spectral multiplexing equipment "T8 Volga" was used, including a

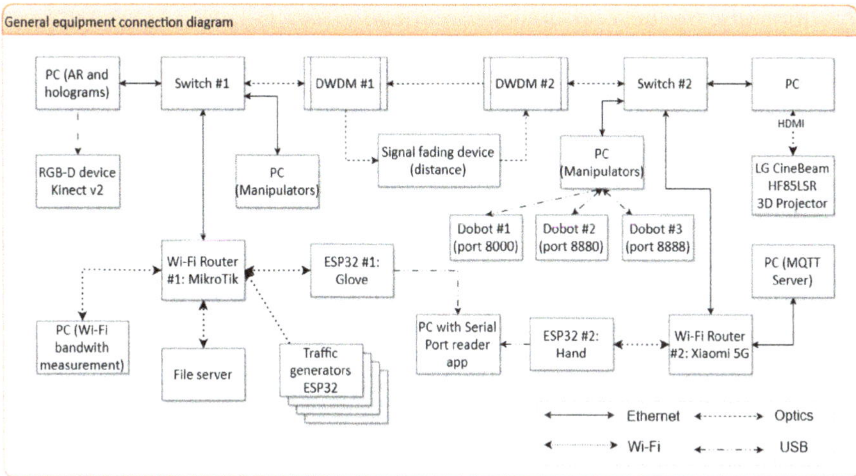

Figure 6.7 General diagram of laboratory equipment connection.

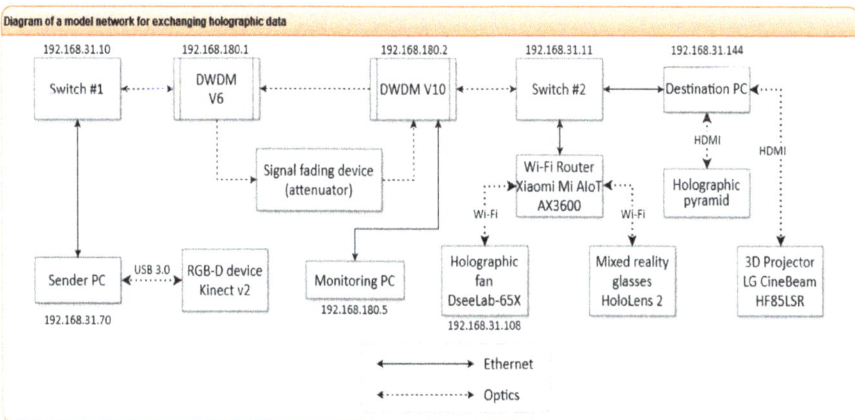

Figure 6.8 A model network for exchanging holographic data.

DWDM V6 multiplexer, which combines information channels into one optical fiber, and a DWDM V10 demultiplexer, which extracts optical signals from the mainstream on the receiving side and distributing them to individual receivers. To create more realistic conditions, a Grandway FHA2S01 attenuator was installed in the main optical line, capable of adjusting the signal attenuation in a wide range from 0 to 80 dB with the ability to select steps of 1 and 0.05 dB, thereby allowing to simulate different lengths of the fiber optic line.

Additionally, the network under test contains two D-Link xStack DES-3526 Layer 2 managed switches with Fast Ethernet support, belonging to the sender and receiver sides, respectively, and used to regulate the throughput of the 100Base-TX ports and subsequently evaluate the impact of these measures on the overall quality holographic image. Network cables are used to connect the sending PC and the receiving PC, as well as patch cords UTP/RJ45 and Cat. 5e.

6.5.1 Scenario for testing holographic traffic streaming to a 3D projector

To organize the streaming of holographic traffic, the Open Broadcaster Software (OBS) Studio version 27.2.3 tool was selected using the Network Device Interface (NDI) plugin, which is a software standard for high-quality transmission and reception of high-definition video in real time over a computer local low latency network. OBS Studio has many advantages, including the ability to capture video from an external RGB-D camera, support for a large number of filters for video sources (including color correction and chromakey identification), as well as high data processing speed, which is achieved through the use of powerful compression algorithms packages that reduce the amount of transmitted traffic. To speed up image transmission, the holographic frame is divided into two parts: moving and stationary. As a result, when streaming, OBS Studio updates only those parts of the image where changes have been detected in real time.

The process of preparing and transferring holographic data to a 3D projector can be divided into several successive stages:

1) Installation of Kinect for Windows SDK 2.0 software on the sending computer.
2) Connecting the Kinect v2 for Xbox One RGB-D camera to a computer via a USB 3.0 interface.
3) Launching the OBS Studio streaming program version 27.2.3.
4) Removing the green background by selecting and configuring the Chroma Key tool in the “Filters” menu.
5) Installing the NDI plugin for OBS Studio.
6) Select the “NDI Output settings” section from the “Tools” drop-down menu and check the box next to “Enable NDI Output” in the window that appears.
7) Connecting a projector with 3D support to the recipient computer – LG CineBeam HF85LSR via the HDMI interface.

8) Install the NDI plugin for OBS Studio on the receiver computer.
9) Adding "NDI Source" to the drop-down menu by clicking the "+" icon at the bottom of the OBS Studio window in the "Sources" tab.
10) Selecting "create a new source" and search in the drop-down list for the name of the computer broadcasting the holographic stream (the name of the sending computer with the prefix "OBS").
11) Playback of a streaming holographic video broadcast received from the sending computer via an NDI connection.

6.5.2 Test scenario for 3D image transmission to a holographic fan

This implementation is aimed at generating an animated 3D model of a person with subsequent transfer to a holographic fan. To create 3D models, an RGB-D (red green blue-depth) device from Microsoft Kinect v2, which scans objects and generates a point cloud as presented in Figure 6.9 [15], is used. This cloud contains the x, y, and z coordinates for each pixel of the frame, as well as information about the intensity of the light and its direction (according to the results of many studies, the holographic model in the future will contain information about space, user position and point of view, as well as about 43 different perspectives for each pixel of the image).

Then, the obtained coordinates are transferred to the "3D Scan" software version 2.0.47.0, where they are correlated, supplemented with missing points

Figure 6.9 Point cloud obtained by scanning a 3D object.

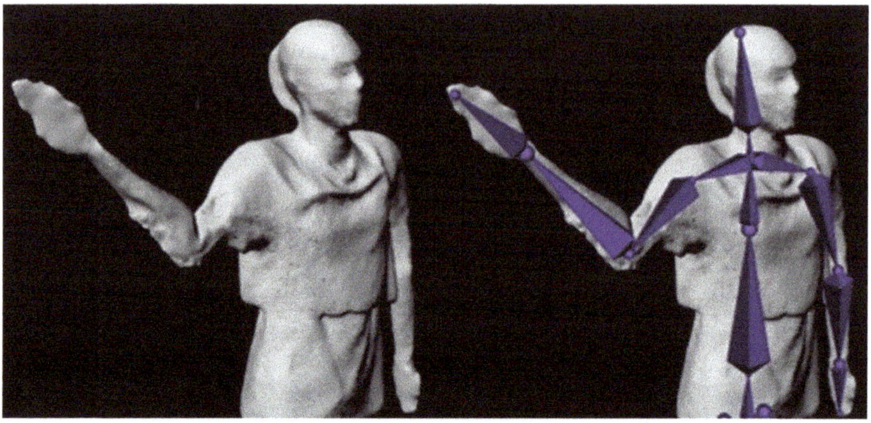

Figure 6.10 Overlaying bone structure on a 3D model.

calculated using mathematical algorithms, and converted into the final 3D model with possible inaccuracies and construction errors. This experiment requires a number of mandatory adjustments for further correct operation of the model, which will be performed in open-source 3D computer graphics software developed by Ton Rosendahl – Blender version 3.3.1.

Due to imperfect shooting conditions (insufficient lighting, changed shooting angles due to the lack of a tripod, inconsistent axis of rotation of the object), the 3D model required editing, including removing unnecessary polygons, connecting adjacent unconnected points, and changing rotation and position. The modification was done manually, without the participation of automatic software. Once edited, a skeletal structure was applied to the model, as presented in Figure 6.10, with each bone requiring individual movement to its intended location, and weight was applied so that when any bone was rotated, its corresponding part of the 3D model would rotate.

To emulate movement, you need to add key frames in extreme positions and specify their number and frequency, thereby determining the duration of the animation. In Blender, you should add lighting elements that will eliminate dark areas and set up a camera that determines the scope and boundaries of the frame. After setting up the export and rendering of the model, a video file is generated for sending to the holographic fan.

The DseeLab-65X display device, presented in Figure 6.11, is an intelligent holographic display that uses continuous viewing (POV) technology based on high-density LED paddles rotating at 750 rpm to produce 3D images with a resolution of 1024 × 1024 pixels and a viewing angle of 160°.

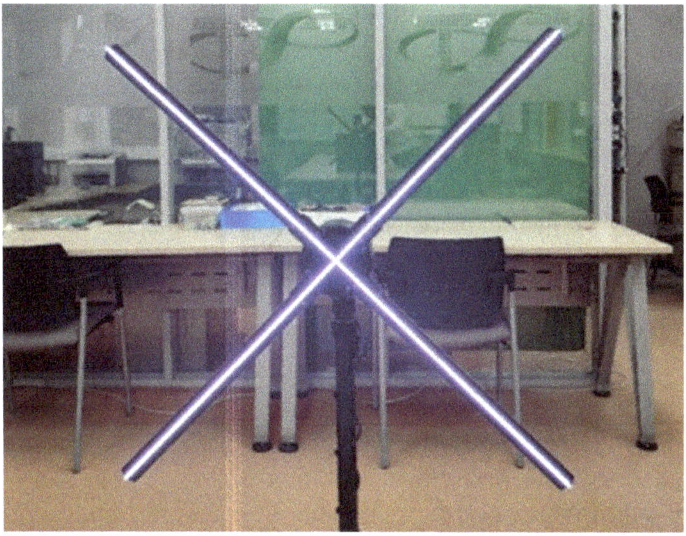

Figure 6.11 DseeLab-65X holographic fan.

The advantages of this device include energy efficiency, portability, satisfactory image quality, and performance. The holographic fan can be controlled by remote control, iOS and Android apps (APP), PC software, and cloud platform, and it can play various image and video stream formats. The process of preparing a holographic fan for transferring three-dimensional data to it can be divided into several successive stages:

1) Connect a mobile device to a Wi-Fi network distributed by a holographic fan (SSID: DseeLab_65X_50011240).
2) Install the DseeLab_Pov application on your smartphone and then launch it.
3) Select the device connection method: "Device in general mode" (the Wi-Fi network of the holographic fan is in the list of SSIDs visible to the smartphone).

Thanks to these actions, it was possible to configure a number of fan parameters, including the order in which the playlist is played, the brightness of the LEDs, and the speed of rotation of the blades. and also provided access to setting a personal identifier and password for the laboratory's local Wi-Fi network to which the holographic device must be connected.

After successfully connecting the holographic fan to the network, the AP/STA LED indicator changes color from green to blue, indicating that the device has been assigned an IP address. As a result, the fan became accessible from any node on the local network, and control was transferred to the desktop version of the DSee.Lab 2.5.8 application. This will make it easier to load animated 3D models onto the fan in the future.

To load a 3D model onto a holographic fan, it must be added to the library (playlist) of the control application, as presented in Figure 6.12. When transferring a holographic object between a user PC and a holographic device, the TCP protocol is used with a preliminary connection establishment, which includes agreeing on the initial number of the packet sequence and other parameters associated with the wireless connection until the process of exchanging traffic via Wi-Fi will begin.

As a result, a 3D model will be displayed on the holographic fan presented in Figure 6.13, which requires adjustment of several basic parameters:

- Blade rotation speed, available in three modes
- Centering the image along the *x* and *y* axes
- Frame tilt angle
- Brightness of the displayed model

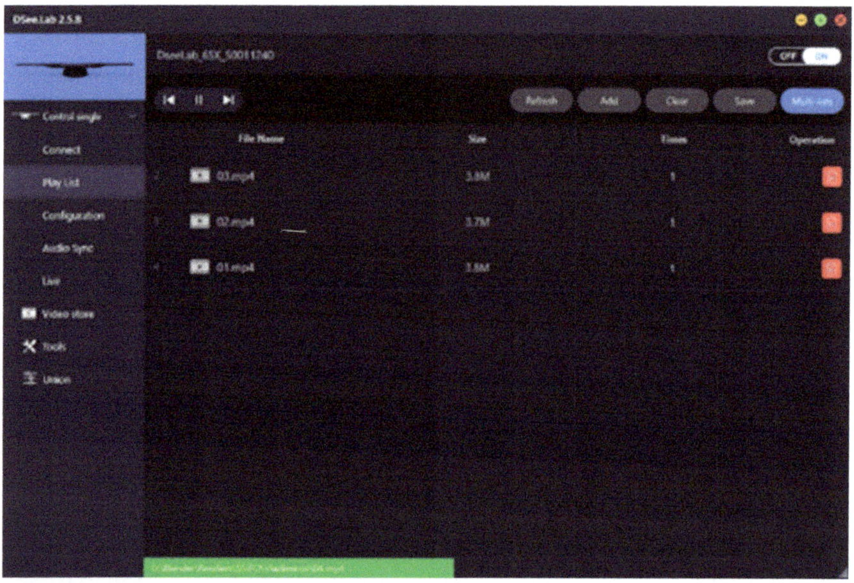

Figure 6.12 Program interface of the desktop version of the application that controls the holographic device.

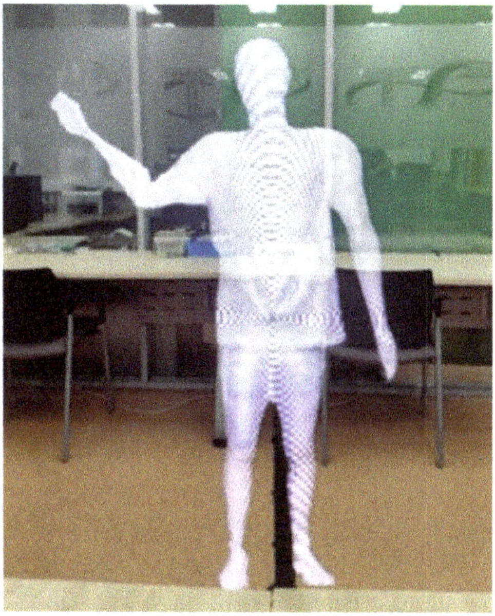

Figure 6.13 3D model displayed on a holographic fan.

6.5.3 Scenario for testing holographic traffic streaming on Microsoft HoloLens 2 mixed-reality glasses

Microsoft HoloLens 2 is a mixed-reality headset that allows users to interact with virtual objects in the real world. They are equipped with high-quality displays that display images on glass in front of the user's eyes, creating the illusion of the presence of virtual objects in the physical world. HoloLens 2 is also equipped with a range of sensors, including cameras, gyroscopes, and accelerometers, which allow them to detect the user's location and orient the image accordingly. The glasses also have a built-in microphone and speaker, which makes it possible to interact with the voice system built into them. With HoloLens 2, users can control virtual objects using hand gestures and voice commands, opening opportunities for learning, design, and entertainment.

Microsoft HoloLens 2 is a powerful tool for HTC, allowing users to communicate in real time using holographic objects, images, and sound. This device can be used to create more interactive and effective forms of communication. The process of preparing and transferring holographic data to Microsoft HoloLens 2 MR glasses can be divided into several successive stages:

1) Launching the OBS Studio streaming program version 27.2.3.
2) Removing the green background by selecting and configuring the Chroma Key tool in the "Filters" menu.
3) Setting up OBS broadcast, as presented in Figure 6.14, in real time via the RTP protocol in unicast format mode: rtp://192.168.31.74:1234, where 192.168.31.74 is the IP address of the receiving device on the network (Microsoft HoloLens 2), 1234 – port (you can specify any free, even number). The MTU limit setting is not required for the RTP protocol. The container format must be set to "rtp_mpegts," the video encoder must be libx264, and the bitrate must be 2500 kbps. The "Video bitrate" field is responsible for the quality of the transmitted video. The standard value "2500 Kbps" is sufficient for high-quality transmission of Full HD 1080p video over a local network; however, depending on the specific task, this parameter can be changed up or down. In the case of an NVIDIA video card installed on the PC that supports hardware H264 conversion, the "libx264" codec can be replaced with "h264_nvenc (libx264)." Set the "Audio Bitrate" field to 96 kbps and select the "aac" codec in the "Audio Encoder" list.
4) Start the broadcast by clicking the "Start Recording" button. After this, the video stream can be viewed on any device (PC, HMD, etc.) connected to the same local network as the computer with OBS Studio broadcasting holographic data.
5) To view the video stream, you need to install a player application on Microsoft HoloLens 2 glasses that can work with streaming video transmitted via RTP. The VLC (VideoLAN Client) media player is great for this task and has versions for various operating systems, including Windows and Android.
6) Launch the VLC application on Microsoft HoloLens 2 glasses and then connect to the broadcast. To do this, you need to select the "network" mode and specify the IP address previously specified in the OBS program in the "enter a URL" field, and also add the "@" symbol. For the unicast mode, the request will look like this: "rtp://@192.168.31.74:1234."

As a result, a holographic image processed by OBS Studio and transmitted via a local network from the RGB-D Kinect camera to MR glasses will be broadcast before the user's eyes in real time. Figure 6.15 presents the real-time holographic broadcast on Microsoft HoloLens glasses.

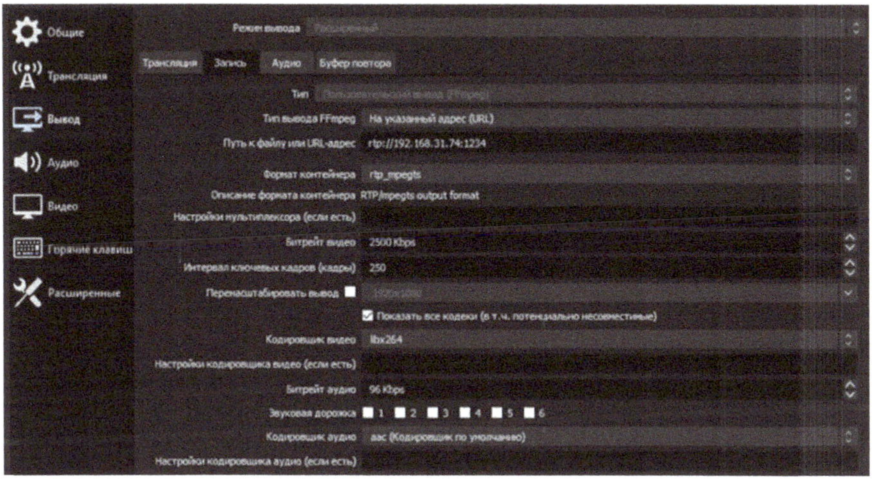

Figure 6.14 Settings for unicast broadcasting via RTP protocol.

Figure 6.15 Real-time holographic broadcast on Microsoft HoloLens glasses.

6.5.4 Scenario for testing holographic traffic streaming to a truncated pyramid

A holographic pyramid is a device of simple design, which is a polyhedron formed by a pyramid and its section parallel to the base. This design creates the illusion of three-dimensionality for the viewer, making the image or video "float" in the air. The truncated pyramid can be made of plastic or glass and works based on the principle of Pepper's Ghost (an illusionistic technique used in theater and magic tricks) [16]. Four symmetrically opposite versions (shifted by 90°) of the same image are projected onto the four sides of the pyramid. In fact, each side projects (refracts) the image falling on it into the center of the pyramid (provided that the edge angle is maintained at 45°). After that, these projections are synchronized, forming a complete figure, and are perceived by the human eye as a single three-dimensional image.

In this study, the holographic pyramid was made of organic glass based on the following ratios: the bottom base is equal to the width of the screen, the top edge is 1:6, and the height is approximately 3:5 the width of the screen.

To create a holographic broadcast in real time in the Open Broadcaster Software (OBS) program, the video stream received from the Kinect v2 camera was prepared by scanning four identical image elements within a square with the green background removed and arranged in a cross-shape, as presented in Figure 6.16.

After that, high-quality transmission of high-definition video and audio streams was organized in real time over a computer local network with low latency from the sending PC to the receiving PC using the NDI (Network Device Interface) plug-in [17].

To monitor and regulate the compliance of the pyramid sizes, the following steps should be performed:

1) Install the prism on the screen, turning its smaller base down and centering the structure according to the special markings (as presented in Figure 6.17).
2) Look at the image above. The size of the smaller square (cut off the top of the pyramid) should be approximately 2 times less than the distance between objects moving on the screen.
3) The image should not extend beyond the larger square.
4) The height of the prism must correspond to the angle of inclination of the rib – approximately 45°. Then, the image will not be too high to go beyond the transparent structure or too low.

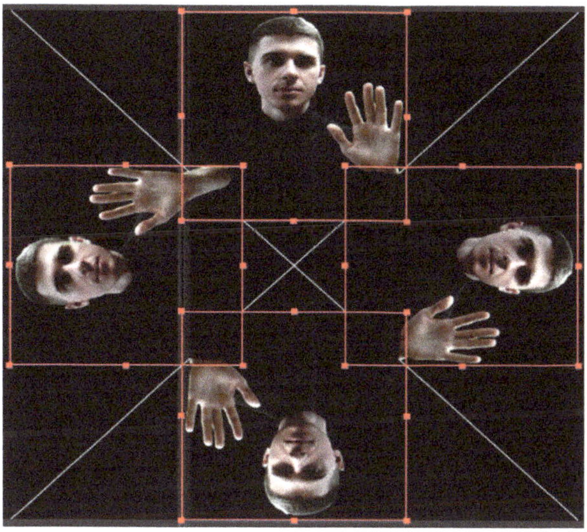

Figure 6.16 Layout of four images on a black background for a holographic pyramid.

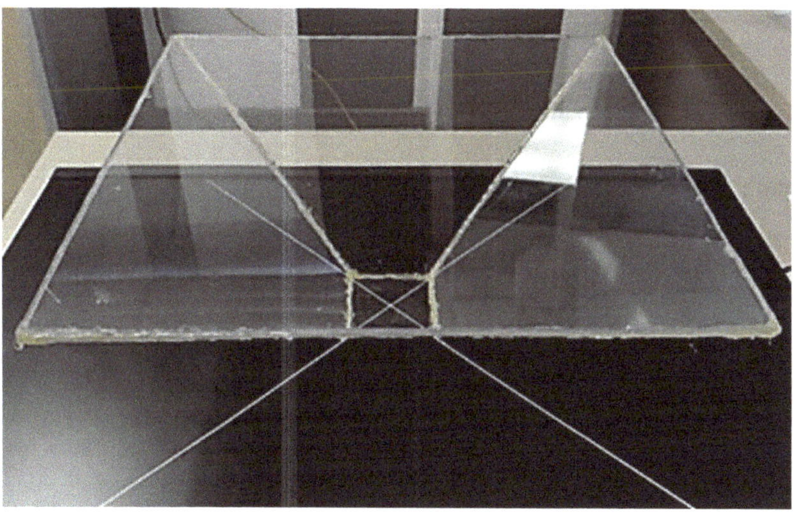

Figure 6.17 Setting the position of the holographic pyramid.

If all conditions meet the requirements, then we can assume that the truncated pyramid is suitable for use in reproducing a volumetric holographic image (as presented in Figure 6.18).

Figure 6.18 Holographic broadcast on a truncated pyramid in real time.

Figure 6.19 Holographic terminal using two projection monitors.

During the research, the design of the holographic stand was refined in order to achieve an increased area of the projection surface while maintaining the width dimensions (as presented in Figure 6.19).

The new concept does not require an equivalent increase in the diagonals of the projecting monitors with an increase in the height of the holographic glass structure (with the same screen diagonal, the height of the model possible for projection doubles; see Figure 6.20).

The terminal includes a transparent stand with mounts for two monitors (bottom and top), as well as a PC connected to the internet for the ability to send and receive holographic traffic.

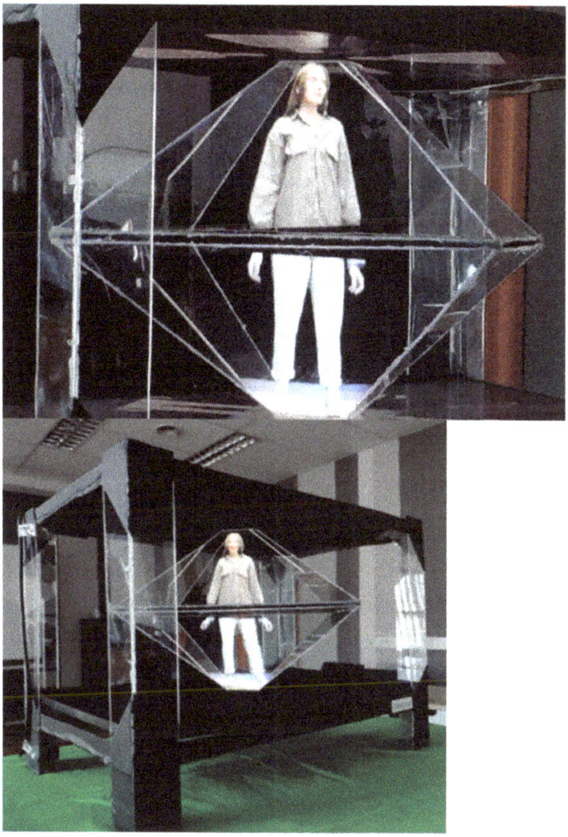

Figure 6.20 Holographic broadcast on the terminal in real time.

The peculiarity of the projection onto a holographic terminal is the need for simultaneous transmission of two video streams obtained when shooting an object with two cameras (conditionally dividing it in half), as well as the requirements for mirror reflection of each of the four images of one stream in relation to the other.

6.6 Summary

In the development of HTC network technologies, it is necessary to take into account high data transfer rates and guaranteed latency. Despite the introduction of new technologies, holographic communication must gradually develop based on existing protocols. In addition, HTC is supported by semantic communication, XR, multimedia, point cloud encoding, and decoding. Solving the problems of streaming HTC content is associated with difficulties that require optimal transport layer protocols. Also important is the introduction of new technologies at the network level, such as New IP and Packet Wash. Wireless sensing can be an important element in the development of HTC, and artificial intelligence can be applied at every level to provide prediction and classification capabilities.

In this chapter, a fragment of a model network has been developed to study the quality of the provision of holographic HTC services. Based on the model network, four scenarios for testing 3D image transmission for various playback devices are proposed. An analysis of technologies and means of streaming holographic images was also carried out, and the values of threshold parameters that directly affect the quality of transmission were determined.

The study was carried out within the laboratory of the Department of Communication Networks and Data at St. Petersburg State University of Technology. The testing included the generation, transmission, and playback of holographic images on four different devices, namely a 3D projector, a holographic fan, mixed-reality glasses, and a holographic frustum.

The study showed that TCP and UDP are currently the main protocols for streaming holographic content. TCP provides a reliable connection but has a higher end-to-end latency than UDP, which does not provide a reliable connection and has a higher packet loss rate. For HTC services, MPEG-DASH and UDP-based protocols such as QUIC and WebRTC are promising technologies, but they require further development and study for HTC data transmission. MPEG-DASH can provide adaptive streaming using HTTP, QUIC can be used to deliver 360° video on XR systems, and WebRTC can provide real-time streaming between browsers or other applications.

In this study, the IP internetworking protocol was used to deliver data in the model network, but in the future, it is proposed to use the New IP concept, which allows packets to be delivered based on service requirements and selectively dropped based on priority values when congestion occurs.

Data transmission at the physical and data link layers faces problems of high latency and low reliability. Terahertz communications, which provide high bandwidth, can provide data rates of several terabits per second and are the core technology for holographic communications. High energy efficiency, reliability, and the coexistence of wireless communication and wireless sensing should be considered when designing communication systems.

References

[1] I. F. Akyildiz and H. Guo, “Holographic-type communication: A new challenge for the next decade,” ITU Journal on Future and Evolving Technologies, vol. 3, no. 2, pp. 421–442, 2022.

[2] A. Ateya *et al.*, “Model mediation to overcome light limitations—toward a secure tactile Internet system,” *J. Sens. Actuator Netw.*, vol. 8, no. 1, p. 6, 2019.

[3] R. Li, “Towards a new internet for the year 2030 and beyond,” in *Proc. 3rd Annu. ITU IMT-2020/5G Workshop Demo Day*, 2018, pp. 1–21.

[4] I. Sodagar, “The MPEG-DASH standard for multimedia streaming over the internet,” *IEEE Multimed.*, vol. 18, no. 4, pp. 62–67, 2011.

[5] S. Zhao, Z. Li, and D. Medhi, “Low delay MPEG DASH streaming over the WebRTC data channel,” in *2016 IEEE International Conference on Multimedia & Expo Workshops (ICMEW)*, 2016.

[6] R. Li, K. Makhijani, and L. Dong, “New IP: A data packet framework to evolve the internetÂă: Invited paper,” in *2020 IEEE 21st International Conference on High Performance Switching and Routing (HPSR)*, 2020.

[7] R. Li *et al.*, “A framework for qualitative communications using big packet protocol,” in *Proceedings of the ACM SIGCOMM 2019 Workshop on Networking for Emerging Applications and Technologies*, 2019.

[8] L. Dong and A. Clemm, “High-precision end-to-end latency guarantees using packet wash,” in *2021 IFIP/IEEE International Symposium on Integrated Network Management (IM)*, IEEE, 2021, pp. 259–267.

[9] FG-NET, “New services and capabilities for network 2030: description, technical gap and performance target analysis,” FG-NET2030 document NET2030-O-027, 2019.

[10] A. Nasrallah *et al.*, “Ultra-Low Latency (ULL) Networks: The IEEE TSN and IETF DetNet Standards and Related 5G ULL Research,” *IEEE Commun. Surv. Tutor.*, vol. 21, no. 1, pp. 88–145, 2019.

[11] J. Farkas, L. L. Bello, and C. Gunther, “Time-Sensitive Networking Standards,” *IEEE Commun. Stand. Mag.*, vol. 2, no. 2, pp. 20–21, 2018.

[12] I. F. Akyildiz, C. Han, Z. Hu, S. Nie, and J. M. Jornet, “Terahertz band communication: An old problem revisited and research directions for the next decade,” *IEEE Trans. Commun.*, vol. 70, no. 6, pp. 4250–4285, 2022.

[13] C. Chaccour, “Building the Foundations and Experiences of 6G and Beyond Networks: A Confluence of THz Systems,” in *Extended Reality (XR), and AI-Native Semantic Communications*, 2023.

[14] C. Liaskos, S. Nie, A. Tsioliaridou, A. Pitsillides, S. Ioannidis, and I. Akyildiz, "A new wireless communication paradigm through software-controlled metasurfaces," *IEEE Commun. Mag.*, vol. 56, no. 9, pp. 162–169, 2018.
[15] C. Cao, M. Preda, and T. Zaharia, "3D point cloud compression: A survey," in *Proceedings of the 24th International Conference on 3D Web Technology*, 2019, pp. 1–9.
[16] B. Christopher, "Explaining the Pepper's Ghost Illusion with Ray Optics," *COMSOL Blog*. [Online]. Available: https://www.comsol.com/blogs/explaining-the-peppers-ghost-illusion-with-ray-optics/. [Accessed: 20-Apr-2024].
[17] "NDI – Technology for data transmission over local networks." [Online]. Available: https://streampark.ru/blog/ndi-tehnologiya-peredachi-dannyh-po-lokalnym-setyam/. [Accessed: 20-Apr-2024].

7

Development of Research Methods for Assessing the Quality of Transmission of Telepresence and Metaverse Data

7.1 Introduction

Today, there are several methods and approaches to assessing the quality of video transmission. They can be divided into two large groups: subjective and objective approaches. Objective approaches are of great interest to operators, developers, and service providers since they allow not only to assess user satisfaction with a service but also to predict this assessment by analyzing the values of network performance characteristics. For holographic playback devices, the difficulty lies in the fact that it is necessary to analyze large volumes of information in a short time, which significantly affects the performance of servers and other equipment. In order to increase the effectiveness of objective assessments, it is necessary to determine the parameters that have the greatest impact on the provision of services and establish their relationship with subjective assessments.

Due to the fact that the holographic stream has a large set of characteristics, at the moment, there is no objective method developed that allows one to definitely say whether the user is satisfied with the HTC service or not. As a result, the focus needs to be on a subjective approach, the concept of which is to assess the quality of the holographic image directly from service users [1]. Visual quality is the characteristic of the transmitted holographic object compared to the original. Such characteristics depend on recording methods, encoding/decoding methods, equipment used, and network parameters.

Quality of experience (QoE) provides telecommunications service providers with insight into how end customers perceive the quality of their services. Factors such as latency, jitter, packet loss, etc., can directly impact the experience quality score [2]. QoE monitoring allows providers to identify problems and take corrective action.

7.2 Analysis of Existing Methods for Subjective Assessment of Video Transmission Quality

According to Recommendation ITU-R BT.500-14 [3], the following methods can be used to conduct subjective assessment sessions:

- Distortion scale (DSIS)
- Continuous quality scale (DSCQS)
- Sources of simultaneous exposure (SDSCE)
- Single source continuous quality assessment (SSCQE)
- Subjective assessment of multimedia video image quality (SAMVIQ)

Table 7.1 provides a comparative description, highlighting assessment session features, structures, and stages [3].

In order to obtain reliable results, it is necessary to involve at least 15 observers. Fewer raters may be used for studies with limited scope, such as experimental studies. In addition, it is important to consider observers' competence level in assessing the quality of a holographic image. Among the existing subjective methods, one can distinguish methods with two sources of influence (DS, double stimulus) and alternative methods. By two sources of influence, it meant that each pair of videos, consisting of a reference and a test video, was played back sequentially during the experiment. Examples of two-source methods are DSIS, DSCQS, and SDSCE methods.

Alternative methods, in turn, are divided into two large groups: methods with one source of influence (SS, single stimulus) and methods of comparing influences (SC, stimulus comparison). In SS methods, only the video sequence under test is played back and evaluated. Impact comparison methods involve playing back a pair of video sequences and assessing quality relative to each other.

Analysis of existing QoE assessment tools of recommendation ITU-R BT.500-14 contributed to the development of a new technique with its own scale of criteria and is suitable for assessing the quality of transmission of holographic images, considering their features [3].

7.3 Development of a Method for Assessing the QoE of Holographic Images

The quality of perception of HTC services is jointly influenced by three main indicators: image quality, depth quality, and visual comfort [4]. Image quality is the characteristic of an image (resolution, clarity, detail, etc.) perceived by

Table 7.1 Subjective methods for assessing images.

Method	**DSIS**	**DSCQS**	**SDSCE**	**SSCQE**	**SAMVIQ**
Demonstration of the undistorted standard	Yes	Yes	Yes	No (viewing conditions close to real)	Constant access to an explicit reference
Notifying the observer about the demonstration of the standard	Yes	No	Yes	No	Yes
Presentation structure	The reference and evaluated images are shown once or twice	A pair of images, consisting of an undistorted and a possibly distorted representation, is shown simultaneously one or more times for equal periods of time	Simultaneous presentation two signals – a reference (original) and a test signal (subject to some distortion)	One-time demonstration sequences	Access multiple sequence options, each shown separately
Duration sequences	∼10 c	∼10 c	Adaptively changing depending on the content of video sequences and the type of errors	∼3–5 min	10–15 c

Table 7.1 *Continued.*

Method	DSIS	DSCQS	SDSCE	SSCQE	SAMVIQ
Assessment session periods	Reference image (gray), evaluated image (gray)	Test sequence A (gray), test sequence B (gray)	Sequence of different PV/UI pairs without separation (FV – fragment of video image UI – test condition)	Sequence of one or more TC/PC combinations without separation (TK – content type, for example, HTC, MR, AR. PC – quality parameters, for example, bit rate)	Access to the content of the next sequence is provided after the viewer evaluates all options
Sequence order	Pseudo-random order in image presentation and distortion	Pseudo-random order using all required combinations	Pseudo-random order	Pseudo-random order	The sequence option is selected randomly by the viewer using a graphical computer interface
Number of playback devices	1	1	1 or 2 nearby	1	1
Grading scale	Five-point distortion rating scale	Double continuous vertical scale divided into five segments	Vertical slider in the range 100–0	Continuous vertical scale divided into five segments	Continuous scale (0–100) with five linear quality ratings
Grading period	Making a decision during the period T_4 = 5–11 seconds	Making a decision during the period T_4 = 5–11 seconds	Continuous grading throughout the demonstration	Continuous grading throughout the demonstration	The ability to review each option any number of times, stop and change your assessment

Table 7.1 *Continued.*

Method	DSIS	DSCQS	SDSCE	SSCQE	SAMVIQ
Interpretation of results	Evaluation of the second sample in relation to the first	Difference between evaluation under reference and test image conditions	Identifying differences between two sequences and assessing the accuracy of video information (perfect or zero)	Comparison of data from all test sessions (graph of average quality score versus time)	Quality assessment performed scene by scene, including explicit reference, hidden reference, and various video processing algorithms
Session duration	No more than half an hour	No more than half an hour	No more than half an hour	From 30 to 60 minutes	Tests end when all sequences are scored

the user and determines the quality of the system. It depends on many factors, including the level of distortion and various artifacts, technical parameters, and errors that may be associated, for example, with encoding or image transmission.

Depth quality is the ability of a system to convey a realistic sense of volume (the accuracy of spatial depth in an image). Linear perspective, blur, gradients, and others create the illusion of depth, even in standard 2D images. However, 3D holographic images also include information about image offset, depth of field, alignment accuracy, and image matching across different viewing angles, which enhances the user's depth perception.

Visual comfort or discomfort is an individual sensation associated with viewing 3D images. Images that are not captured or displayed correctly can cause serious discomfort and fatigue to the viewer. Some researchers argue that the psychological impact of holographic imaging technologies can be assessed in terms of broader concepts such as naturalness and presence.

The naturalness of perception of a holographic image is determined by its ability to convey reality with a high degree of realism (plausibility). Various distortions in the image can significantly reduce its naturalness. For example, objects in a 3D image may appear unnaturally large or small (puppet effect) or unnaturally thin (cardboard effect). Presence can be defined as the subjective feeling of being in one environment or situation when one is actually in another. This is often called "immersiveness," which is the feeling of immersion in a virtual or artificial environment.

The samples being evaluated are sequences of holographic video images subject to the influence of algorithms with various methods of introducing attenuation and errors into communication channels, unique technologies for encoding and transmitting images, variable speeds of switching equipment interfaces, various transmission scenarios, etc. Using this method can also be useful for comparing different codecs, resolutions, and bitrates in holographic videos, as well as comparing different image processing techniques such as noise reduction and sharpening.

7.3.1 Holographic QoE rating scale

In the process of testing the concept of the HTC service, a 10-point scale was developed that considers indicators of image quality, depth, visual comfort, and criteria for assessing the holographic type of communication. These indicators allow for obtaining more accurate QoE and recording minor impacts on image parameters. These indicators are as follows.

1) **Rating 10:**
    - Excellent quality of the holographic video image (comparable to the original).
    - The image color is saturated.
    - Full detail.
    - Realistic sense of volume.
    - A stereoscopic effect is observed.
    - The image is highly natural.
    - The sense of presence (immersiveness) is fully achieved.
    - The video material starts instantly (there is no delay in the video stream).
    - The frame rate is high.
    - The image moves smoothly.
    - Without fading or delays.
    - There is no lag in the audio stream.
    - No excessive noise in the image.
    - The viewer does not experience discomfort or fatigue.
2) **Rating 9:**
    - Excellent quality of the holographic video image (comparable to the original).
    - The image color is saturated.
    - Full detail.
    - Realistic sense of volume.
    - A stereoscopic effect is observed.
    - The image is highly natural.
    - The feeling of presence is fully achieved.
    - The video material on display starts instantly (there is no delay in the video stream).
    - The frame rate is high.
    - The image moves smoothly but sometimes fluctuates slightly.
    - No lag in the audio stream.
    - No excessive noise in the image.
    - The viewer does not experience discomfort or fatigue.
3) **Rating 8:**
    - Excellent quality of the holographic video image (comparable to the original).
    - The image color is saturated.

- Full detail.
- Realistic sense of volume.
- A stereoscopic effect is observed.
- The image is highly natural.
- The feeling of presence is fully achieved.
- The video material on display starts instantly (there is no delay in the video stream).
- The frame rate is high.
- The image moves smoothly.
- There is a slight delay.
- Sometimes, the video stream freezes.
- No delay in the audio stream.
- No excessive noise in the image.
- The viewer does not experience discomfort or fatigue.

4) **Rating 7:**

- Excellent quality of the holographic video image (comparable to the original).
- The image color is saturated.
- Full detail.
- Realistic sense of volume.
- A stereoscopic effect is observed.
- The image is highly natural.
- The feeling of presence is fully achieved.
- The video material on display starts instantly (there is no delay in the video stream).
- The frame rate is average.
- The image moves smoothly, but slight fluctuations are noticeable.
- There is a delay of about 1 second between the movement of an object in front of the RGB-D camera and its display on the holographic device.
- Sometimes, the video stream freezes.
- No lag in the audio stream.
- No excessive noise in the image.
- Discomfort and fatigue are not observed in the viewer.

5) **Rating 6:**

- Good quality of the holographic video image.
- The image color is saturated.

- Partial detailing.
- An acceptable sense of volume.
- A stereoscopic effect is observed.
- The image is natural.
- A sense of presence is achieved.
- The video material on display starts after a certain period.
- The frame rate is average.
- The image moves smoothly, but small fluctuations are noticeable.
- There is a delay of about 1 second.
- Sometimes, the video stream freezes.
- There is no delay in the audio stream.
- There is no excessive noise in the image.
- The viewer does not experience discomfort or fatigue.

6) **Rating 5:**

- Good quality of the holographic video image.
- The image color is saturated.
- Partial detailing.
- An acceptable sense of volume.
- A stereoscopic effect is observed.
- Good naturalness of the image.
- A sense of presence is achieved to an average extent.
- The video material on display starts after a certain period.
- The frame rate is average.
- Noticeably, there are gaps in the frames of the stream.
- The image moves intermittently.
- Vibrations are noticeable.
- A delay of about 2 seconds is observed.
- Sometimes, the video stream freezes.
- There is no delay in the audio stream.
- There is no excessive noise in the image.
- There is minor discomfort and fatigue for the viewer.

7) **Rating 4:**

- Holographic video image with minor artifacts.
- The image color is desaturated.
- Partial detailing.
- Insufficient sense of volume.
- There is a stereoscopic effect.
- Reduced naturalness of the image.

- A sense of presence is achieved to an average extent.
- There is a long delay between the start of the video broadcast and the image's appearance on the screen.
- Frequency frames are average.
- There are noticeable gaps in the frames of the stream.
- The image moves intermittently.
- Fluctuations are noticeable.
- There is a delay of about 2 seconds.
- The video stream often freezes.
- There is a delay in the audio stream.
- There is no excessive noise in the image.
- There is minor discomfort and fatigue for the viewer.

8) **Rating 3:**

- Holographic video image with visible artifacts.
- Image color is desaturated.
- Partial detail.
- No sense of volume and stereoscopic effect.
- Reduced naturalness of the image.
- A sense of presence is not achieved.
- There is a long delay between the start of the video broadcast and the appearance of the image on the screen.
- The frame rate is low.
- Gaps in the sequence of frames of the stream are very noticeable.
- The image moves intermittently.
- There are distinct fluctuations in the image.
- There is a delay of about 3 seconds.
- The video stream often freezes.
- There is a delay in the audio stream.
- There is excessive noise in the image.
- Significant discomfort and fatigue are observed in the viewer.

9) **Rating 2:**

- Holographic video image with severe and frequent damage.
- No sense of volume and stereoscopic effect.
- Extremely low naturalness of the image.
- A sense of presence is not achieved.
- Many frames are lost, objects may be indistinguishable.
- Difficult to perceive.

- The image color is dull.
- Partial detailing.
- There is a long delay between the start of the video broadcast and the appearance of the image on the screen.
- The frame rate is very low.
- Gaps and lost frames of the stream are very noticeable.
- The image moves intermittently.
- The delay between frames is about 4 seconds.
- Distinct fluctuations in the image are observed.
- Fading of the video stream appears after each frame.
- There is excessive noise in the image.
- Severe discomfort and fatigue in the viewer are observed.

10) Rating 1:

- A holographic video image in which nothing is visible or understandable.
- There is no sense of volume and stereoscopic effect.
- The naturalness of the image and the feeling of presence are not achieved.
- Most frames are lost.
- The color of the image is dull.
- Partial detail.
- There is a significant delay between the start of the video broadcast and the image's appearance on the screen.
- The frame rate is extremely low.
- Gaps and lost frames make it impossible to perceive the video stream.
- The image either moves intermittently with a frequency of 1 frame every 5–7 seconds or freezes completely, and the broadcast stops.
- Severe discomfort and fatigue are observed in the viewer.

7.3.2 Test organization

Assessment of the quality of perception of holographic images should be carried out cyclically and is based on comparing the original reference sequence of frames with the sequence in which certain distortions were introduced, covering possible combinations. Using this method and the developed scale, it is possible to evaluate the impact of different types of distortions on the properties of 3D video and determine which distortions lead to large losses in quality. The undistorted (original) and distorted video streams must be presented in a random order, and their sequence must be different for each viewing session, covering all possible combinations (Figure 7.1). The assessment session is limited to half an hour. As a result of the tests, the average score is calculated for each video sequence and test condition. Each session includes a specific order as follows.

1. Reference image – T_1 = 10 seconds.
2. Gray background – T_2 = 3 seconds.
3. Evaluated image – T_3 = 10 seconds.
4. Time to make a decision – T_4 = 5–11 seconds.

Experience from existing studies shows that increasing the T_1 and T_3 periods beyond 10 seconds does not improve the ability of raters to evaluate image sequences correctly. Participants must take part in test sessions of the following format:

- **Content fragment (FC):** Each FC corresponds to a specific type of content (for example, holographic broadcast, demonstration of 3D objects, AR video) and is processed using one of the tested quality parameters (QP) (e.g., bit rate).
- **Test session (TS):** This is a sequence of one or more FC/PC combinations in random order. Each SI must include all FCs and PCs, but not necessarily all FC/PC combinations.
- **Test demonstration (DT):** This is a complete test of all FCs and PCs.

The viewing distance is selected depending on the type of holographic device and the size of the image it projects. Table 7.2 shows the optimal parameters for each of the tested holographic display systems. It is also necessary to limit the maximum viewing angle relative to the perpendicular

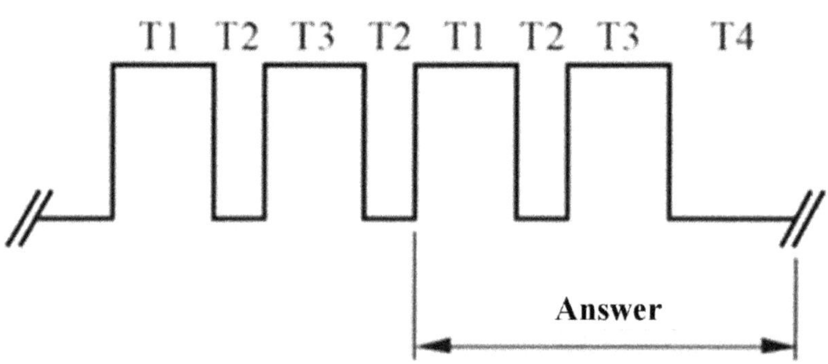

Figure 7.1 Structure of presentation of materials for testing.

Table 7.2 Information on optimal viewing distances in image heights (H) and required horizontal angles.

Device	Image resolution	Reference document	Screen format	Optimal horizontal viewing angle	Optimal viewing distance
3D projector	1920 × 1080	ITU-R BT.709 [5]	16:9	32°	3,2 H
Holographic fan	1024 × 1024	XGA [6]	1:1	15°	4,3 H
MR glasses	2556 × 1704	ITU-R BT.2020 [7]	3:2	Fixed at 76°	Fixed
Holographic pyramid	Equivalent to screen resolution	–	1:1	21°	3,5 H

so that changes in color rendition that occur during playback on devices are not noticeable to the viewer.

7.3.3 Structure of the assessment session system for the holographic pyramid

To assess the quality of the image broadcast on the holographic terminal (double truncated pyramid), signals must be sent through a software switch (OBS Studio functionality [8]), which is controlled by the session administrator. The signal can be an image supplied directly from the source (Kinect v2), a uniform gray background that separates the stages of presentation, or an image that has undergone a processing process (change in resolution, frequency, bitrate, etc.). The optical attenuator allows you to introduce distortion into the video stream by attenuating the signal level (the administrator sets various attenuation values ranging from 0 to 27 dB).

Evaluators are presented with a series of test images or sequences that are paired together. In this case, the first image in the pair follows a direct path from the source (attenuation is 0 dB), and the second image is the same copy but has already passed through a signal attenuation device or processing process (Figure 7.2). The duration of the session should be no more than 30 minutes. To establish consistency among observers, it is recommended that approximately five "practice demonstrations" be conducted at the beginning of the first session.

7.3.3.1 Factors influencing the QoE assessment of an image broadcast on a holographic pyramid

The Dual Frustum Holographic Pyramid is a device used to reproduce holographic images and allows the viewer to view holography without using

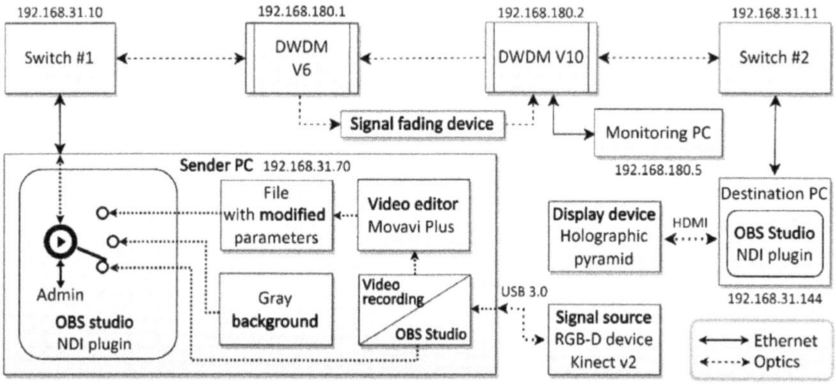

Figure 7.2 General layout of the system for conducting QoE assessment sessions for a holographic pyramid.

special glasses. However, the quality of the image directly depends on the illumination and is sensitive to external factors.

Light can affect the image produced using a holographic frustum in several ways:

1) A light source's brightness and color temperature can change the perception of color and its intensity. For example, bright light can cause stray reflections on the pyramid's surface, thereby deleting the image quality. In such cases, using the pyramid under controlled lighting is recommended.
2) The location of the light source relative to the pyramid can influence the required angle and optimal viewing distance for the viewer. For example, if the light source is too close to the pyramid, the observer needs to change his position in order to prevent the visibility of rays reflected from the glass surfaces and exposing the broadcast image. In such cases, it is necessary to place the pyramid in a place with minimal illumination and exclude direct rays of light.
3) The size of the light source affects the amount of unnecessary surface reflections. In other words, the smaller the light source, the fewer reflections, and shadows there will be on the pyramid's surface, which improves the quality of perception of holographic objects.

So, when using a truncated holographic pyramid, it is recommended to maintain controlled lighting conditions, exclude direct light rays, and use a light source of minimal size to obtain the best image quality. When designing a holographic pyramid, it is also important to maintain certain proportions and dimensions to ensure high-quality image reproduction. This is due to the fact that the holographic pyramid uses the principles of reflection and refraction of light from the basic laws of geometric optics, which creates the impression of three-dimensionality of the image.

Failure to comply with proportions and dimensions can lead to image distortion or complete loss. In addition, when designing a holographic pyramid, it is necessary to consider the angles of glass (45°), which are key elements. If the angles of the faces do not meet the requirements, this can lead to a shift in perspective or deformation of objects. Failure to comply with the angles of the glass can also lead to a loss of brightness and contrast in the image and result in a "shaking" effect. The consequences result from the lack of complementarity of perpendicular reflections at the central points of the pyramid and the appearance of the phenomenon of pictures shifted relative to each other.

In general, maintaining proportions and dimensions when designing a holographic pyramid and maintaining the correct angles of glass are key factors in ensuring high-quality perception and increasing the subjective assessment of viewers.

7.3.4 Structure of the evaluation system for the holographic fan

To assess the quality of the image broadcast on a holographic fan, the testing scheme must include an image forming device, Kinect v2, a sending computer with installed video capture and editing software, OBS Studio and Movavi Video Editor Plus, as well as the Dsee application. Lab POV [9] allows you to control a holographic fan (Figure 7.3). The session administrator creates three different video files, as follows.

1. The first one is the original image obtained from the RGB-D camera.
2. The second one is an image with converted parameters (changed resolution, frequency, intensity, bitrate, color saturation, added noise, jitter defects, etc.).
3. The third is a uniform gray background.

Next, the created files are uploaded to the playlist of the playback device and presented to the evaluators as a series of images or videos.

7.3.4.1 Factors influencing the QoE assessment of the image broadcast by a holographic fan

According to the developed methodology, the holographic fan must be rigidly mounted on a telescopic tripod at a distance from observers approximately 4.5 times the height of the broadcast image. Due to the specific design of

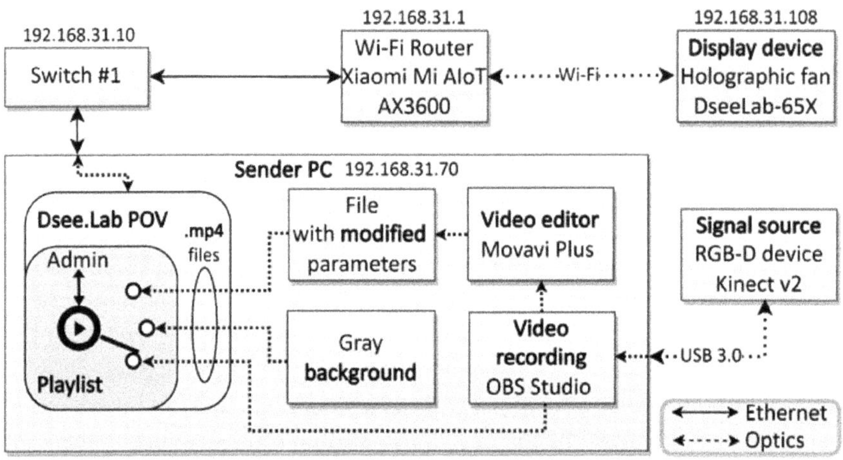

Figure 7.3 General layout of the system for conducting QoE assessment sessions for a holographic fan.

the playback device and the insufficient density of LEDs on the blades, the problem of low detail and unclear frames arises. In addition, limitations in processing power cause delays and image synchronization artifacts, which reduce the quality of the experience. Finally, blade vibration and noise from blade rotation also have a negative impact on QoE.

The study revealed that the quality of perception of a three-dimensional image is significantly influenced by such characteristics as the frame rate and the density of RGB diodes on the blades of a holographic fan. Challenges were also identified in the optimal layout of fan hardware components to achieve maximum processing power in a compact package with an efficient cooling system.

7.3.5 Structure of the evaluation system for MR glasses

To assess the quality of the image transmitted on MR glasses, the testing scheme includes an image forming device, Kinect v2, a sending computer with installed video capture and editing software, OBS Studio and Movavi Video Editor Plus, as well as the VideoLAN Client application for Windows Universal Platform (VLC) [10], which allows MR glasses to receive holographic traffic (Figure 7.4). During the evaluation session, three different video streams are generated, as follows.

1. The first is the original image obtained from the RGB-D camera.
2. The second is an image that has undergone processing.

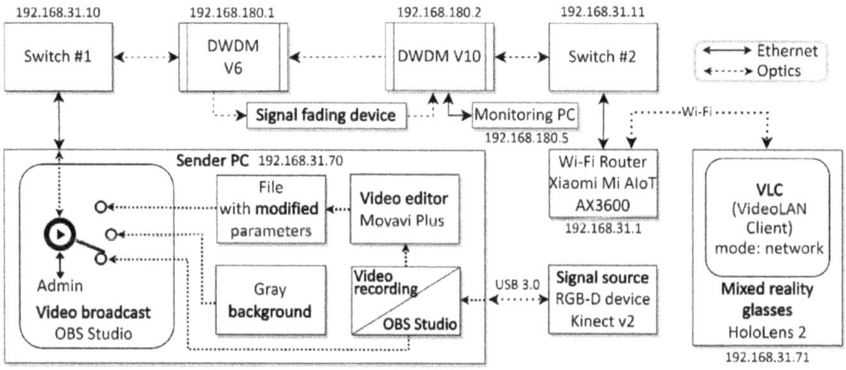

Figure 7.4 General layout of the system for conducting QoE assessment sessions for MR glasses.

3. The third is a uniform gray background.

7.3.5.1 Factors influencing the QoE assessment of a holographic image broadcast on MR glasses

In the modern world, the main condition for creating three-dimensional images is the presence of at least two views of the same scene, shot with two cameras located on the same horizontal axis. As a result, objects in images will have different relative positions for the left and right eyes. This difference in the position of objects between two frames is often called image disparity (or parallax) and can be expressed in pixels, physical units (such as mm), or relative values (such as percentage of screen width).

Parallax in 3D is an effect in which objects in the foreground appear closer than objects in the background and move faster based on the viewer's head movement. This creates a deeper, more realistic sense of depth and perspective in 3D images. However, using this effect in MR glasses during testing caused discomfort and eye fatigue for some observers. This is due to the fact that the eyes and brain must constantly adapt to changing depth and perspective of the image, which contributes to the tension of the eye muscles, and the proximity of the glass's lenses to the eyes of the observer also has a negative impact.

Additionally, if the parallax effect is used in holographic images that are constantly moving, such as in real-time video streams or when viewing objects in 3D, the discomfort is greatly amplified and may result in headaches. The movement of images in augmented reality causes a phenomenon known as "train motion" in a person, that is, the feeling that the body is in motion and moving, although it is in a static position. In addition to the above, the glass wearer may feel pressure on the face caused by the device fitting tightly and touching the bridge of the nose, which similarly contributes to a poor user experience.

To reduce the risk of parallax fatigue in holographic images transmitted to MR glasses, it is necessary to use 3D technologies with a moderate level of offset from one image to another. Also, you should not look at the 3D image for long without breaks, but you should pause regularly and rest your eyes. In addition, you can fine-tune the parameters of the three-dimensional image to make it smoother and more comfortable for perception.

7.3.6 Structure of the assessment session system for multimedia holographic content

When organizing tests of multimedia content (the presence of an audio stream broadcast synchronously with a holographic video image), the following requirements must be considered:

1) The test is carried out in stages in accordance with Figure 7.5.
2) The ability to play and evaluate holographic video sequences in any order is available for each scene. Each sequence can be played and scored several times.
3) Access to video material during the transition between scenes occurs randomly to prevent viewers from voting in accordance with the established order.
4) To move to the next scene, you must rate all sequences of the current scene.

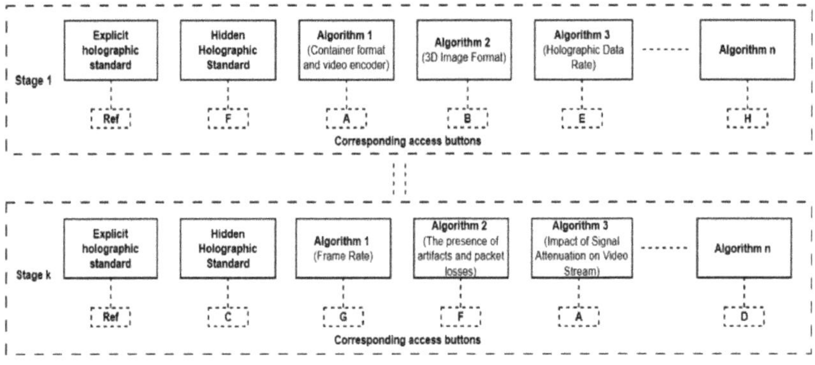

Figure 7.5 An example of organizing a holographic broadcast test using the SAMVIQ method.

5) To complete the test, you must evaluate all sequences of each scene.

When assessing the quality of service using audio, observers should be informed that they must evaluate not only the quality of the holographic video image but also the overall quality. The viewer randomly selects several options for holographic sequences from a graphical computer interface. After that, he can review each option as many times as he wants, stop and change his assessment. The developed method includes several sequence variants, which can contain both raw and processed sequences, including a reference. Each sequence option is demonstrated separately and assessed using a 10-point scale.

Quality assessment is carried out for each scene separately, including an explicit reference, a hidden reference, and various video processing algorithms. An explicit reference is an original or source sequence that serves as a quality standard for comparison with other sequences.

A hidden master is an alternate version of the original sequence that contains changes, such as compression or processing but is still high quality. The latent standard can test raters' ability to distinguish quality between the original sequence and the modified version. This method is effective for assessing the quality of multimedia content since it considers various characteristics of holographic image processing, such as codec type, image format, data transfer rate, frame refresh rate, presence of artifacts, etc. Each of these characteristics or a combination of them is called an algorithm. An interval of 10–15 seconds is sufficient viewing time for each sequence to obtain an accurate and reliable quality assessment.

7.4 Automation of the Method for Evaluating Holographic Images using the Telegram Bot API

In the course of studying the impact of factors related to the quality, depth, and visual comfort of technologies for transmitting three-dimensional objects, an automated system for assessing the QoE of holographic images was developed based on the Telegram chatbot with support for tested playback devices: a holographic terminal (truncated pyramid), a holographic fan, and MR-points.

7.4.1 Programming interfaces and libraries used

Telegram uses its own encryption protocol; MTProto API (aka Telegram API) is the API through which the Telegram application communicates with the server [11]. To write bots, there is an add-on to the Telegram API called the Telegram Bot API, which uses the HTTPS protocol to exchange data with Telegram servers [12]. Developers can use the API to create bots in various programming languages, including Python, Java, JavaScript, Ruby, PHP, and more. The bot being developed must be able to send requests to the Telegram server and receive updates from it, which is certainly more convenient to do with the help of libraries.

According to statistics, Aiogram is considered the best library for writing bots in Python. It is asynchronous, uses decorators, contains convenient development tools, and provides a powerful interface for interacting with the Telegram API. The advantages of Aiogram include the following criteria [13]:

- Ease of use: Aiogram has a clean, simple, and intuitive API that simplifies creating bots.
- Multi-threading support: Aiogram supports running multiple functions in the background, making it fast and efficient.
- Asynchronous work: Aiogram can work asynchronously with the Telegram Bot API, which allows you to create bots that quickly and efficiently process many requests.
- Full support for all functions of the Telegram Bot API: Aiogram provides full support for all Telegram Bot API functions, allowing you to create bots with any functionality.
- Convenient and flexible event processing mechanism: Aiogram allows you to define handlers for any events that may occur in the Telegram bot, such as a new message, a new chat participant, etc.
- Easily customizable keyboards and menus: Aiogram provides the ability to create easily customizable keyboards and menus that help users interact with the bot.
- Ability to work with databases: Aiogram provides the ability to work with various databases, such as SQLite, MySQL, PostgreSQL, etc., which allows you to store and process large amounts of data.
- An open and actively developing project.

The main components of Aiogram are:

- Dispatcher – handler of incoming messages and events.
- Bot – interface for interacting with the Telegram Bot API.
- Types – module containing all types of Telegram API objects used in Aiogram.

PyCharm was used as an integrated development environment (IDE), which provides a wide range of tools for writing, debugging and testing Python applications [14].

7.4.2 Creating a Telegram bot

To authenticate a bot in the Telegram API, a unique identifier is used – Token. It allows the bot to access various Telegram API functions such as sending messages, receiving updates, managing subscribers, etc. To receive a unique token, you need to find BotFather in Telegram, enter the /start command and follow the steps below:

1) Create a new bot by sending the /newbot command.
2) Give the bot a name and select a unique username (username), which will end in “bot” (Figure 7.6).

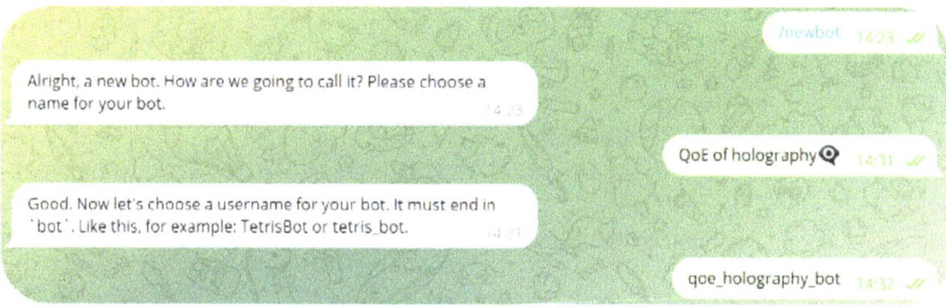

Figure 7.6 Creating and assigning a name to a Telegram bot.

3) After successfully creating a bot, BotFather will send a token, which is necessary to interact with the Telegram API (Figure 7.7).

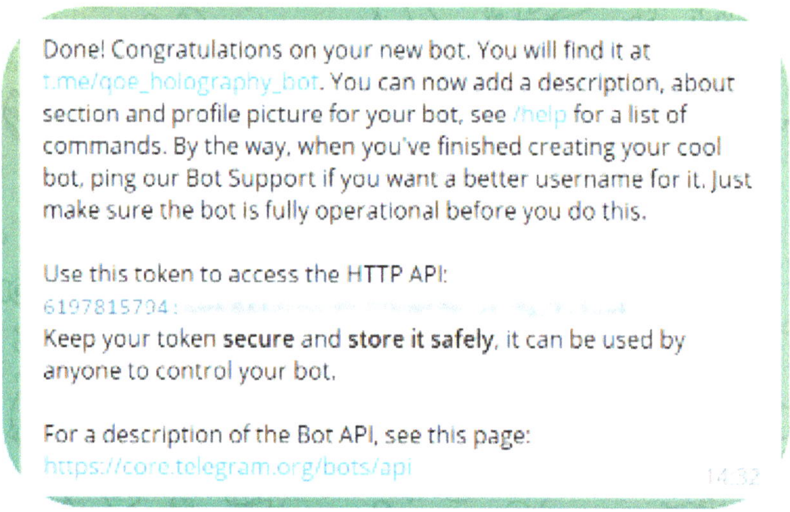

Figure 7.7 Obtaining a token to control the bot.

4) As a result of setting up the bot's profile, its display will look like this (Figure 7.8).

Figure 7.8 Profile of the created bot.

7.4.3 Observer side functionality

When the Telegram bot is initially launched, the appraiser is given the opportunity to check color perception using the Ishihara method (Figure 7.9). To do this, the server sends the observer image No. 1, which shows a number. If the response message is sent correctly, the system sends a similar picture No. 2 to confirm the results.

The next stage is the collection and subsequent entry into the database of information about each of the observers, in order to obtain the most accurate and corrected QoE estimates (Figure 7.10). Recommended details that can be provided include occupational status (e.g., broadcaster, university student, employee), gender, and age category.

It is recommended that raters carefully review the assessment methodology, including possible types of distortions or qualitative characteristics, the rating scale, and the sequence and order of the session. This is done using training sequences that demonstrate the range and type of distortions to be assessed.

A lightweight embedded cross-platform database was used as a database, SQLite, which does not require a separate database management server and provides a simple and fast way to store and access information [15]. A

Figure 7.9 System for checking color perception of appraisers.

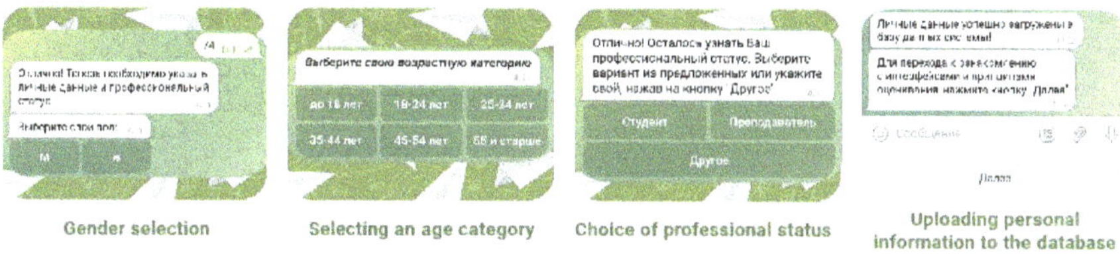

Figure 7.10 Collection of personal information about observers.

distinctive feature of SQLite is its embeddability in applications. This means that most DBMSs are independent applications, interaction with which is organized on the client-server principle. The client program sends a query in SQL language, the DBMS, which can also be located on a remote computer, and returns the query result.

Several libraries are available for working with SQLite in Python, including the built-in sqlite3 library. This library makes it easy to create, modify, and manage databases inside Python scripts. In Figure 7.11, a table from the created database containing information about two appraisers registered

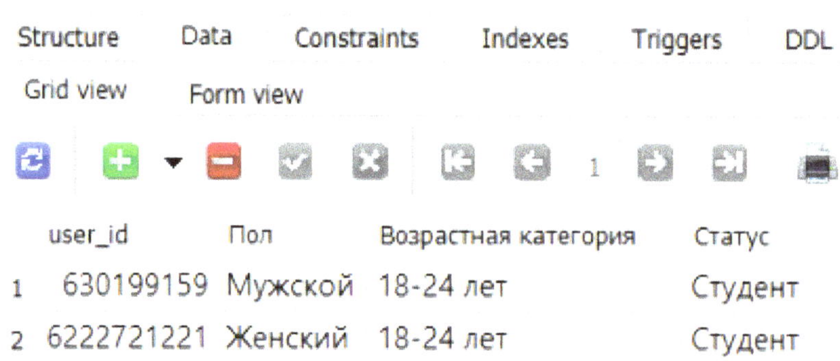

	user_id	Пол	Возрастная категория	Статус
1	630199159	Мужской	18-24 лет	Студент
2	6222721221	Женский	18-24 лет	Студент

Figure 7.11 Table in the database containing personal information about observers.

in the system is shown. User identification in the code is organized on the principle of tracking user_id, due to the fact that it never changes and is assigned uniquely to each Telegram user.

After registering in the system, the user is introduced to the interface and the principle of assessing the quality of holographic images. The next stage is a demonstration session, accompanied by familiarization of participants with various types of distortions, as well as voting procedures using an interactive 10-point rating scale (Figure 7.12).

The developed system is adapted for three reproducing devices: 1) holographic fan; 2) MR glasses; 3) truncated holographic pyramid. Depending on the choice of a particular device, the observer is presented with a unique list of available assessment methods. For each of the methods, the observer can receive a brief information sheet describing the principle of operation and the features of the sequence of the dismantled holographic material (Figure 7.13).

After the user selects one of the proposed playback devices, as well as the QoE assessment method available for it, the system provides visual information about the required viewing distance, indicating the optimal horizontal angle (Figure 7.14).

If the required distance from the observer to the device is successfully established, the evaluator confirms his readiness for the session and goes into the mode of waiting for further instructions from the administrator (Figure 7.15).

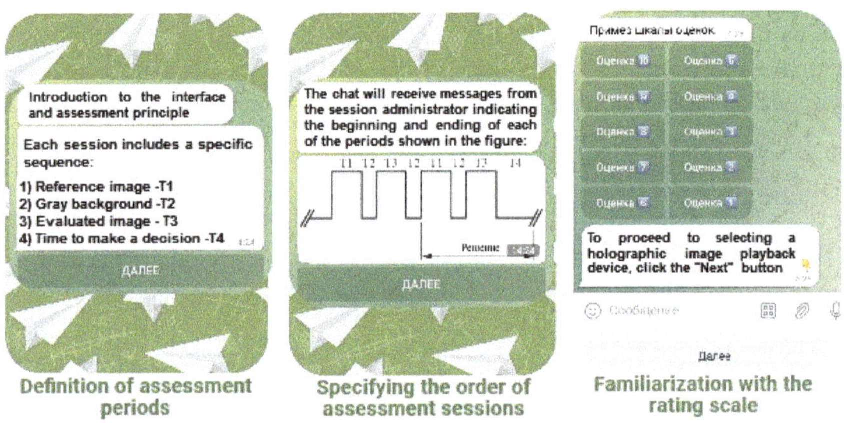

Figure 7.12 Familiarization of observers with interfaces and principles for assessing QoE of holographic images.

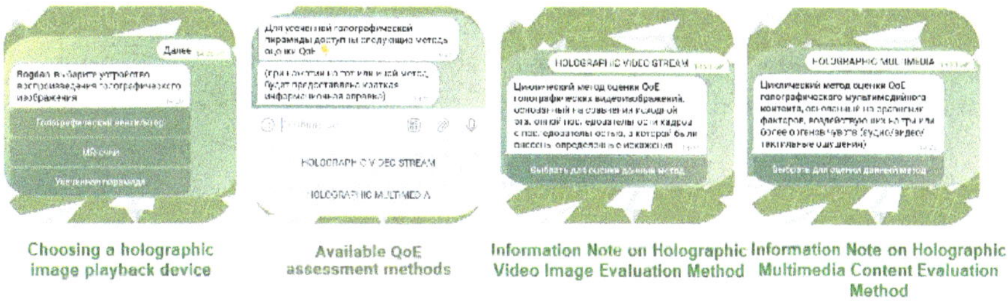

Figure 7.13 Providing available evaluation methods for the selected holographic rendering device.

Figure 7.14 Information about the required distance between the device and the observer.

Figure 7.15 Waiting for further instructions from the session administrator.

7.4.4 Session administrator side

The developed system includes a separate control interface for the session administrator, which is called with the /admin command. The bot also monitors the status of user activity and updates it in the database. Using the "List of online participants" command, the Telegram bot accesses the SQL database and sequentially displays in the chat all rows of the table that satisfy the "active" criterion (Figure 7.16). In addition, the session administrator has the ability to remove a user from the table, thereby limiting their ability to participate in grading.

The developed system implements the ability to simultaneously broadcast messages to all session participants registered in the database. To do this, the administrator needs to select the menu item "Send message" and enter the required text (Figure 7.17). When the message is successfully delivered to users, the session administrator receives a confirmation notification.

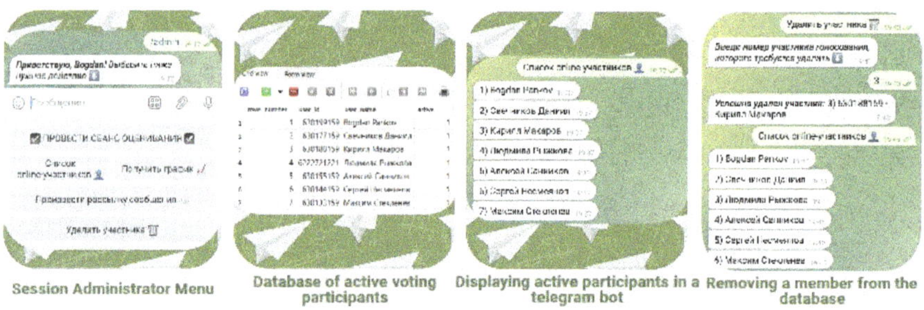

Figure 7.16 Session administrator interface.

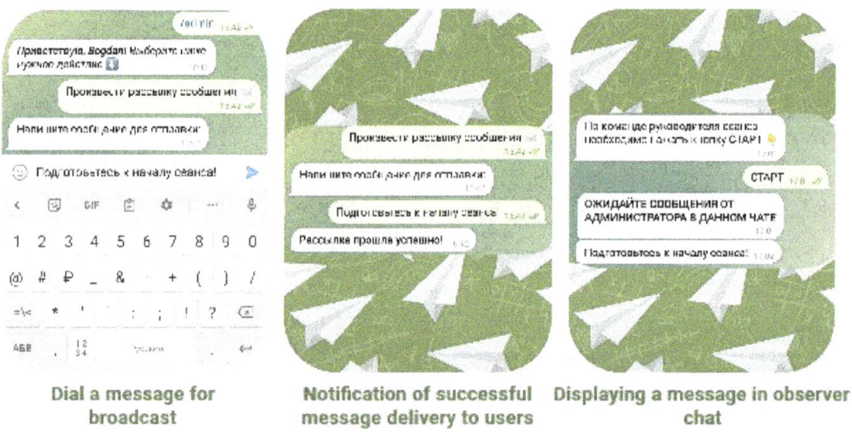

Figure 7.17 Message broadcast functionality.

The main functionality of the system for conducting sessions for assessing the quality of perception of holographic images is available to the administrator after clicking the "Conduct an assessment session" button. After this, you are given the opportunity to select a holographic image display device and a QoE assessment method available for a given device. Confirmation of the choice, in turn, is accompanied by a brief information note, and the administrator is prompted to move to the section with control commands (Figure 7.18).

The "Session started" and "Session stopped" buttons are displayed on the client side in the "Waiting for a message from the administrator" mode (Figure 7.17) and are used to broadcast notification of voting participants about the status of the demo session. The developed system provides for broadcasting a 10-point distortion rating scale presented in Figure 7.19. The observer views the image being assessed for 10 seconds, after which he is given time to make a decision and select an assessment from the proposed list.

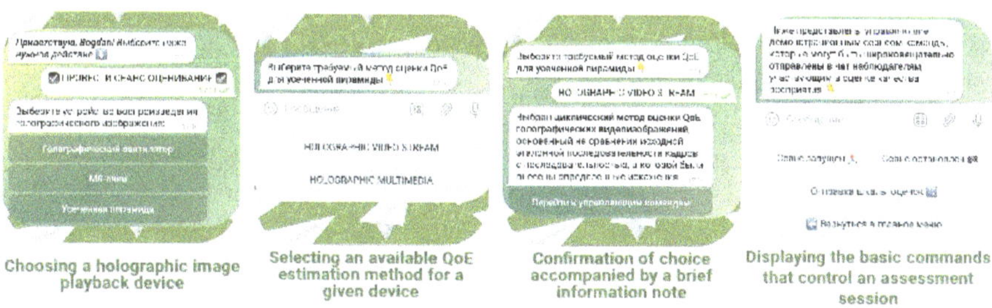

Figure 7.18 The main functionality of the system administrator for conducting QoE assessment sessions of holographic images.

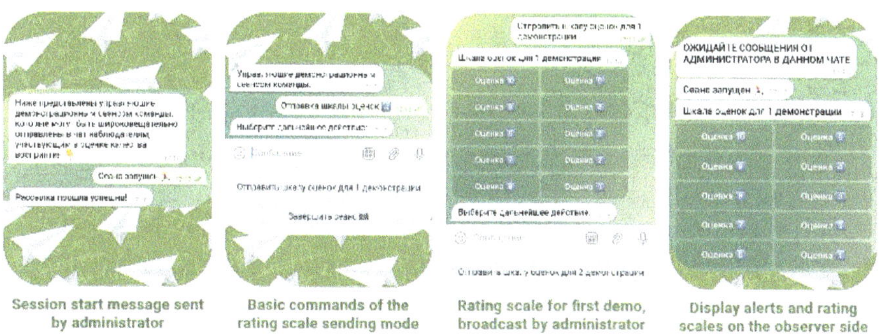

Figure 7.19 Broadcast rating scale during demo session.

Figure 7.20 Recording observer scores.

After the observer clicks on one of the ratings presented in the scale (Figure 7.20), the bot sends a text notification about the successful registration of the choice and enters the information into the corresponding database table (Figure 7.21). Each unique user assessment (user_id) is accompanied by relevant information about the holographic image display device (device), the selected QoE assessment method (method), as well as a demo number (demo_number) and a short description (description).

user_id	device	method	demo_number	assessment	description
6222721221	Усеченная пирамида	HOLOGRAPHIC VIDEO STREAM	1	10	Excellent
6222721221	Усеченная пирамида	HOLOGRAPHIC VIDEO STREAM	2	5	Satisfactory
6222721221	Усеченная пирамида	HOLOGRAPHIC VIDEO STREAM	3	1	Fail
630177159	Усеченная пирамида	HOLOGRAPHIC VIDEO STREAM	1	8	Very good
630177159	Усеченная пирамида	HOLOGRAPHIC VIDEO STREAM	2	4	Satisfactory
630177159	Усеченная пирамида	HOLOGRAPHIC VIDEO STREAM	3	2	Fail

Figure 7.21 Database table containing QoE scores for each demo session and corresponding scoring method.

7.4.5 Analysis of the results of assessment sessions

The system also implements the ability to obtain rating graphs for a given observer in real time (Figure 7.22). To do this, the administrator must select the “Get graph” menu item, and specify the device for reproducing the holographic image and the evaluation method for which information should be displayed. After that, the bot checks the information against the database (Figure 7.21) and displays in the chat a list of observers who took part in voting using the specified method.

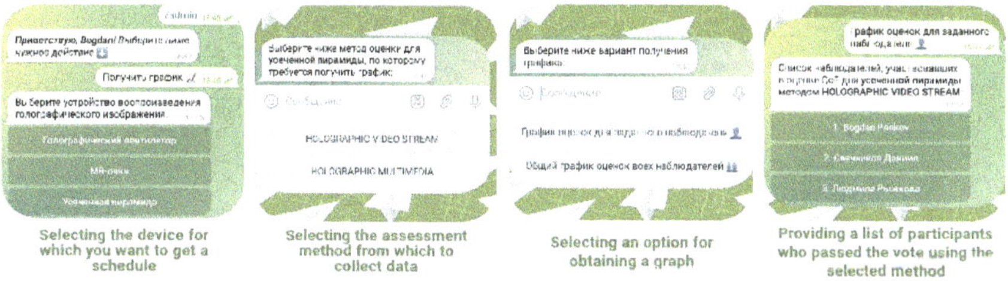

Figure 7.22 Selecting a voter for whose ratings you want to get a graph.

After clicking on one of the proposed participants, the system automatically builds a graph, indicating the demonstration number and observer ratings on the axes (Figure 7.23).

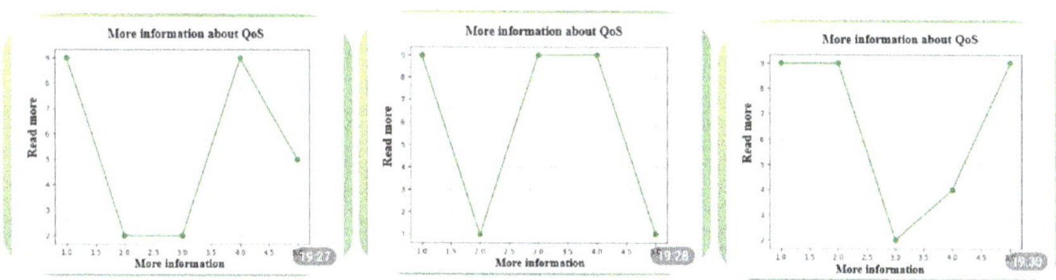

Figure 7.23 Score plots for given observers.

The system also allows you to obtain a common rating schedule for all observers. To do this, you need to click the button of the same name in the administrator menu (Figure 7.24).

7.5 Testing the Developed Method for Assessing the QoE of Holographic Images

In the first stage, an experiment was conducted to study the influence of the level of signal attenuation (measured in decibels) in an optical cable on the subjective assessment of a three-dimensional image (Table 7.3) reproduced on a holographic double pyramid. The evaluation criteria were a developed 10-point scale.

The purpose of assessing the impact of attenuation on traffic was to identify the maximum possible distance over which holographic communications services can be provided while meeting quality of service and perceptual requirements, without the use of signal recovery tools and additional error detection and correction devices.

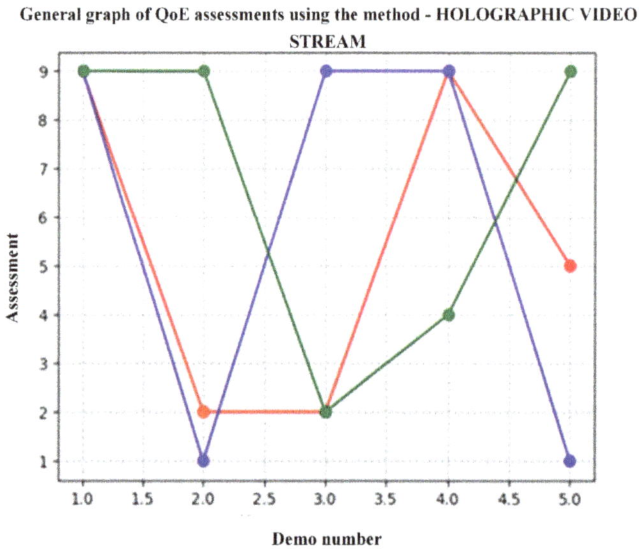

Figure 7.24 General schedule of assessments of all observers who participated in voting using the selected method.

The attenuation value of 27 dB is critical for the simulated holographic communication network, since the number of bit errors per unit time reaches a limit value and the DWDM system goes into emergency mode, stopping traffic transmission.

During the study of holographic transmission, it was possible to achieve acceptable quality of perception (Figure 7.25) at a boundary not exceeding a signal attenuation of 18 dB, which corresponds to the equivalent length of an optical line of 90 km. There was excellent holographic video quality, rich color reproduction, high frame rates, smooth motion, full detail, no audio delay or unwanted interference, and an imperceptible delay between the movement of an object in front of the RGB-D camera and its display on the holographic terminal.

However, with a further increase in the attenuation value in the communication line, various artifacts and damage begin to appear on the holographic video image, detail decreases, color becomes less saturated, and a long delay appears between the start of the video broadcast and the display of the image on the screen. In addition, the frame rate decreases exponentially, gaps in the sequence of frames of the stream become more noticeable, intermittency and distinct fluctuations are observed when the image moves, freezing of

Table 7.3 Influence of signal attenuation on the subjective assessment of the quality of the holographic image.

Signal attenuation (dB) in an optical cable	**Attenuation-equivalent length of optical line (km)**	**Bit error rate (BER) – number of bit errors per unit time**	**Round trip time (RTT) (ms)**	**Subjective assessment of the quality of the reproduced holographic video stream**
0,0	< 1	0,0	0,095	10
2,0	10	0,0	0,154	10
4,0	20	0,0 Hardware error	0,121	9
6,0	30	0,0 correction	0,215	8
8,0	40	0,0 (forward	0,232	7
10,0	50	0,0 error correction	0,398	7
12,0	60	0,0 – FEC)	0,403	7
14,0	70	0,0	0,389	6
16,0	80	0,0	0,467	6
18,0	90	0,0	0,455	5
20,0	100	$7{,}84 \times 10^{-9}$	0,812	4
22,0	110	$4{,}71 \times 10^{-6}$	0,765	3
24,0	120	$1{,}77 \times 10^{-4}$	0,836	3
25,0	125	$7{,}12 \times 10–^{4}$	1,112	2
26,0	130	$1{,}56 \times 10^{-3}$	1,063	2
27,0	135	– (BER threshold exceeded)	– (Connection lost)	1

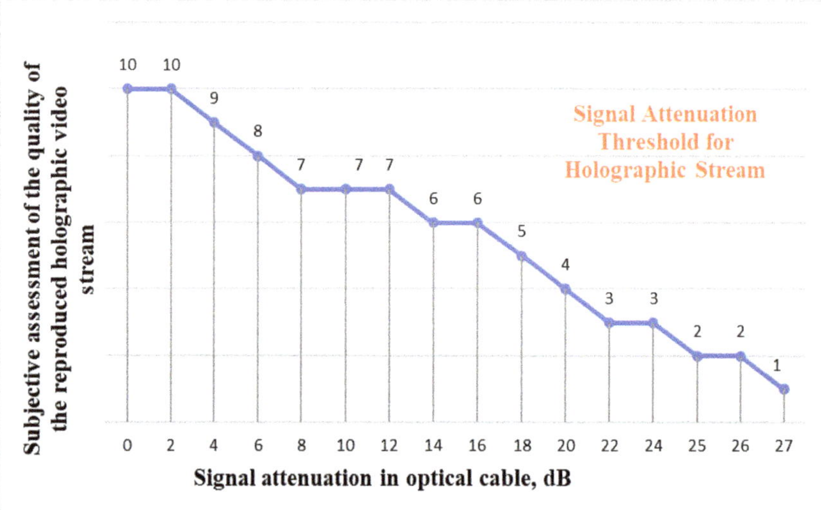

Figure 7.25 Relationship between subjective assessment of the quality of the holographic flow and attenuation in the optical cable.

the video and audio stream often occurs, as well as excessive noise in the image.

To check the impact of bandwidth (in Mbit/s) on the quality of the broadcast holographic video image (Table 7.4), a PC was connected to one of the free RJ45 ports on switch #2 on the traffic recipient side. This made it possible to configure and debug the configuration parameters of the device, sequentially changing the maximum data transmission and reception speed for the selected interface in the "Port Bandwidth" window.

As a result of the experiment, it was found that for successful transmission of a holographic stream from two RGB-D cameras to a truncated pyramid with a round-trip delay (RTT) of <1 ms, the communication channel throughput value should be about 448 Mbit/s. While the minimum throughput (Figure 7.26), at which a sharp decrease in frame rate is observed, an increase in the delay between frames of a holographic video stream, and intermittency and fluctuations in image movement are recorded, significantly worsening perception, is 60 Mbit/s.

The transmission of holographic traffic in the study was organized based on the QUIC (Quick UDP Internet Connections) protocol developed by Google [16] to improve the performance of data delivery in networks. QUIC provides the ability to simultaneously transmit multiple streams of information between two network nodes, operates on top of the UDP (User

Table 7.4 Impact of switch port bandwidth on subjective assessment.

Switch port bandwidth, Mbit/s	Subjective assessment of the quality of a holographic video stream
1000	10
800	10
500	10
200	9
150	9
100	8
90	8
80	8
70	8
65	7
60 (threshold value of bandwidth, below which the frame rate drops sharply, high delays appear between frames of the video stream, and the image begins to move intermittently, with fluctuations)	7
55	6
50	5
40	4
30	4
20	3
15	2
12	2
11	1
10	The image freezes completely and the broadcast stops

Datagram Protocol), provides encryption comparable to TLS and SSL, has lower connection and transmission latency than TCP, and also has the ability to direct (forward error correction) at the packet level.

During the research, the applicability of the developed 10-point scale was also checked to assess the impact of various video stream parameters on the QoE indicator (Figure 7.27). For this purpose, an experimental evaluation session was conducted, allowing the involvement of less than 15 observers. A group of nine evaluators recorded the impact of dynamically changing video stream settings on the final quality of perception of a holographic image (Table 7.5) reproduced on a double truncated pyramid (Figure 7.20).

The resolution and frame rate were changed, container formats such as MPEG-TS (MPEG Transport Stream), RTP-MPEGTS (Real-time Transport

Table 7.5 Influence of video stream parameters on subjective assessment.

Observer number		**1**	**2**	**3**	**4**	**5**	**6**	**7**	**8**	**9**	**Average**
	1080	9	8	9	10	7	8	9	7	9	8,44
Permission	720	8	9	7	7	8	7	7	8	7	7,56
	480	6	4	5	6	4	5	6	4	6	5,11
	60	9	10	8	9	10	9	8	9	8	8,89
Frame frequency	30	8	7	8	7	8	9	9	8	9	8,11
	24	8	8	7	8	6	8	9	8	9	7,89
	MPEG-TS	7	6	6	7	5	6	7	5	7	6,22
Container format	RTP-MPEG-TS	8	7	8	8	7	8	7	8	7	7,56
	HLS	9	9	8	9	9	8	9	9	9	8,78
	9500 Kbit/s	9	8	9	10	9	10	9	9	10	9,22
Bitrate	4500 Kbit/s	8	8	9	8	7	9	7	8	8	8,00
	1500 Kbit/s	5	6	4	3	5	6	4	5	4	4,67
	LibX264	8	8	8	7	8	9	7	8	8	7,89
Video encoder	NVidia-NVENC	9	8	8	9	8	8	9	8	9	8,44
	HEVC	9	8	9	10	8	9	9	10	9	9,00
	BT.2100	9	9	9	9	10	8	10	9	9	9,11
Color space	BT.709	9	8	8	9	9	9	8	9	9	8,67
	BT.601	8	7	6	7	7	6	7	8	6	6,89

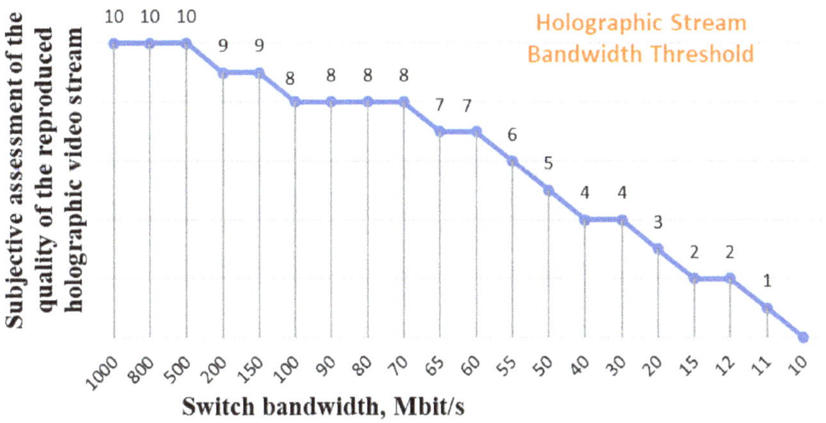

Figure 7.26 Relationship between subjective assessment of the quality of the holographic stream and the bandwidth of the switch port.

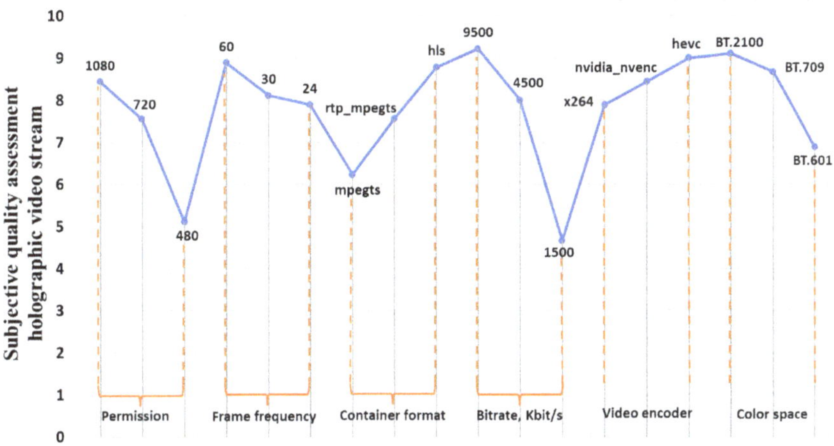

Figure 7.27 The relationship between subjective quality assessment and holographic streaming settings.

Protocol MPEG Transport Stream), and HLS (HTTP Live Streaming) were used, providing mechanisms for data fragmentation for transmission via UDP, RTP, and HTTP protocols, respectively.

Similarly, the impact of the bitrate (the number of bits used to transmit/process data per unit of time) on the holographic Full-HD stream was recorded. Bitrate affects the resolution, detail, smoothness of motion, and overall quality of the 3D video being played back. At low bitrates, the image may appear pixelated and blurry. A high bitrate, in turn, provides a clearer and more detailed image, but requires adequate network bandwidth.

In a similar way, video encoders (data compression and recovery algorithms) changed dynamically, where:

- Software (×264): Encoder that uses the CPU processor.
- Hardware (NVIDIA NVENC or H.264): Uses the GPU video processor (available only for NVidia video cards and reduces the load on the central processor).
- Software (HEVC or H.265): Video compression format using more efficient algorithms compared to H.264 (provides 25%–50% higher compression efficiency with the same video quality).
- The final parameters used to influence the holographic flow were the BT.2100, BT.709, and BT.601 color space standards, of which:
    - BT.2100 [17] defines the transmission and presentation parameters of HDR images, including a wide range of brightness and color depth. BT.2100 provides a more realistic and contrasting video display with improved detail in dark and light areas.
    - BT.709 [18] is the primary standard for color space and color information formats in most high-definition television (HDTV) and standard-definition television (SDTV) systems.
    - BT.601 [19] is a limited range color space standard used in the analog and digital video industries for standard-definition (SD) video.

Considering the above, we can conclude that in order to achieve the desired network parameters of holographic communication, it is necessary to improve algorithms for encoding and processing information, use high-speed transport protocols, as well as update methods for converting and compressing holographic images.

7.6 Directions of Research and Possible Solutions for the Implementation of HTC Systems with a High QoE Indicator

Traditional multimedia communication buffers transmitted or received data to meet network latency and jitter requirements. However, HTC requires high data transfer rates (up to several Tbit/s) and unprecedentedly large resources for storing information, which cannot be supported by existing systems with limited memory capacity. HTC systems face the following major challenges [20, 21]:

- Develop effective methods for encoding and decoding holograms. Directly sending uncompressed holographic data can generate enormous network traffic, which will proportionally increase end-to-end latency and packet loss rates. However, there is a trade-off between encoding/decoding efficiency and computational latency, that is, high encoding speed results in long computational latency, and vice versa. Co-designing optimal encoding and decoding methods based on real-time network conditions and user quality of experience (QoE) requirements is a challenging problem.
- Delivery of data packets with guaranteed and limited end-to-end delay, which can be determined automatically or by the user. Since buffering HTC data is difficult due to the large size of the information, it is desirable to deliver HTC packets at a predetermined time in a deterministic manner. In other words, multiple HTC data streams can be delivered synchronously. Existing networks use best-effort delivery, where the latency probability density function has a long tail, and users cannot adjust latency metrics. HTC networks must allow applications/clients to define end-to-end delay parameters and deliver data packets with a guaranteed and bounded value.
- Accurate synchronization of multi-sensory data. HTC systems use influences on several human sensory organs, which requires synchronization of a certain number of initial sensors, displays, speakers, and actuators. Any asynchronous performance can significantly reduce QoE, but synchronization accuracy in turn also increases end-to-end latency, the amount of memory required, and the complexity of the required computations.
- Development of robust and intelligent rendering algorithms in the presence of packet losses and errors. Packet losses and errors in traditional networks can result in the need for retransmission, which in turn increases end-to-end latency. To avoid this problem, retransmission of information packets is not allowed in HTC networks, and intelligent error correction must be organized on the traffic receiving side in order to ensure high quality of service.
- Support for highly interactive applications. Highly interactive HTC applications typically require more than 60 frames per second when capturing user movements. A single holographic frame has a significant

size, as a result of which a high frame rate can significantly increase the required data transfer rate and network throughput. The applications in question require low end-to-end latency (on the order of 1 ms), and with current networks exceeding 1 ms and encoding and decoding latency can reach several hundred milliseconds, supporting highly interactive HTC applications remains a pressing challenge.

To date, the problems that were analyzed cannot be solved using existing methods. For example, ultra-reliable low latency communication (uRLLC) can provide high reliability and low latency of 1 ms, but the communication throughput cannot reach several Gbit/s to Tbit/s [22]. The HTC's design will involve many aspects of technical innovation that are unavailable today.

7.7 Summary

This chapter describes a subjective approach to assessing the quality of holographic images, which is based on user reactions to the transmitted content. Testing of methods can be carried out both under conditions of ideal system operation and under degraded conditions associated with data transmission or processing. Also, for holographic broadcasting, information was generated on the optimal viewing distances in image heights (*H*) and the required horizontal angles, adapted for various devices for reproducing holographic images.

The study presents several methods that can be used to assess the perceptual quality of 3D images. Each method has its own characteristics and depends on the playback device used. In addition, another method is mentioned – subjective assessment of multimedia video image quality (SAMVIQ).

The study of HTC service perception quality can be determined by three main indicators: image quality, depth quality, and visual comfort. Image quality is subject to technical parameters and errors, and depth quality can create the illusion of volume in 3D holographic images. Visual comfort or discomfort is related to the individual experience of viewing 3D images and can be assessed in terms of naturalness and presence. Distortions in the image can disrupt the naturalness of perception, and a sense of presence can be achieved through immersion.

In the process of testing the concept of the HTC service, alternative criteria for assessing the holographic type of communication were also identified, presented in the form of a 10-point scale indicating the level of achievement of realism and a sense of presence, as well as the influence of delays, noise, frame rate, smoothness of movements on the discomfort, and fatigue of the viewer.

HTC systems require new approaches to multimedia communications, as they have high requirements for data transfer speed and information storage. To achieve optimal performance in HTC systems, it is necessary to solve the following problems: development of effective methods for encoding/decoding holograms, delivery of data packets with guaranteed and limited delay, accurate data synchronization, and development of stable and intelligent rendering algorithms in the presence of packet losses and errors.

References

[1] "Video quality assessment. Subjective assessment of video quality," *IXBT*. [Online]. Available: https://www.ixbt.com/divideo/estimate1.shtml. [Accessed: 02-Apr-2024].

[2] G. Pan, S. Xu, S. Zhang, X. Chen, and Y. Sun, "Quality of experience oriented cross-layer optimization for real-time XR video transmission," *IEEE Trans. Circuits Syst. Video Technol.*, vol. PP, no. 99, pp. 1–1, 2024.

[3] ITU-R, "BT.500-14-Methodologies for the subjective assessment of the quality of television images," ITU, 2019.

[4] S. Ibuki, T. Okamoto, T. Fujihashi, T. Koike-Akino, and T. Watanabe, "Rateless deep graph joint source channel coding for holographic-type communication," in *GLOBECOM 2023 - 2023 IEEE Global Communications Conference*, 2023, pp. 3330–3335.

[5] "BT.709: Parameter values for the HDTV standards for production and international programme exchange," *ITU*, 2015. [Online]. Available: https://www.itu.int/rec/R-REC-BT.709. [Accessed: 20-Jan-2024].

[6] "Extended Graphics Array," *Wikipedia*. [Online]. Available: https://en.wikipedia.org/wiki/Extended_Graphics_Array. [Accessed: 20-Mar-2024].

[7] "BT.2020: Parameter values for ultra-high definition television systems for production and international programme exchange," *ITU*, 2015. [Online]. Available: https://www.itu.int/rec/R-REC-BT.2020. [Accessed: 10-Feb-2024].

[8] "How to use OBS Studio," *OBS Studio*. [Online]. Available: https://obsproject.ru/kak-pol-zovat-sya-obs-studio/. [Accessed: 01-Apr-2024].

[9] "Dsee.Lab POV." [Online]. Available: https://www.dseelab.com/home. [Accessed: 25-Mar-2024].

[10] "VLC media player." [Online]. Available: https://www.videolan.org/. [Accessed: 01-Mar-2024].

[11] "MTProto Mobile Protocol," *Telegram.* [Online]. Available: https://core.telegram.org/mtproto. [Accessed: 01-Apr-2024].

[12] "Telegram Bot API," *Telegram.* [Online]. Available: https://core.telegram.org/bots/api. [Accessed: 15-Apr-2024].

[13] "Getting to know aiogram." [Online]. Available: https://mastergroosha.github.io/aiogram-3-guide/quickstart/. [Accessed: 01-Apr-2024].

[14] "PyCharm," *jetbrains.* [Online]. Available: https://www.jetbrains.com/pycharm/. [Accessed: 25-Mar-2024].

[15] "SQLite." [Online]. Available: https://blog.skillfactory.ru/glossary/sqlite/. [Accessed: 01-Apr-2024].

[16] "QUIC," *Google.* [Online]. Available: https://peering.google.com/#/learn-more/quic. [Accessed: 01-Apr-2024].

[17] "BT.2100: Image parameter values for high dynamic range television for use in production and international programme exchange," *ITU.* [Online]. Available: https://www.itu.int/rec/R-REC-BT.2100. [Accessed: 12-Apr-2024].

[18] "BT.709: Parameter values for the HDTV standards for production and international programme exchange," *ITU.* [Online]. Available: https://www.itu.int/rec/R-REC-BT.709. [Accessed: 12-Apr-2024].

[19] "BT.601: Studio encoding parameters of digital television for standard 4:3 and wide screen 16:9 aspect ratios," *ITU.* [Online]. Available: https://www.itu.int/rec/R-REC-BT.601. [Accessed: 01-Apr-2024].

[20] I. F. Akyildiz, "Keynote speech 3 NETWORKING 2030: Metaverse, extended reality, hologram type and mulsemedia communication challenges for 6G/beyond wireless systems," in *2023 Eighth International Conference on Fog and Mobile Edge Computing (FMEC)*, 2023, pp. 5–5.

[21] H. Zhang, H. Zhang, B. Di, and L. Song, "Holographic integrated sensing and communications: Principles, technology, and implementation," *IEEE Commun. Mag.*, vol. 61, no. 5, pp. 83–89, 2023.

[22] M. Osama, A. A. Ateya, S. Ahmed Elsaid, and A. Muthanna, "Ultra-reliable low-latency communications: Unmanned aerial vehicles assisted systems," *Information (Basel)*, vol. 13, no. 9, p. 430, 2022.

8

6G Telepresence Architecture for Immersive AR/VR and Holographic Experiences: A Layered Approach to Superior QoE

8.1 Introduction

In the increasingly interconnected world, innovative technologies are reaching beyond the limits of physical presence. Telepresence services have emerged as a revolutionary solution, changing how individuals and organizations collaborate, interact, and experience remote environments [1]. Based on the convergence of telecommunications, robotics, virtual reality, and artificial intelligence, telepresence services enable users to bridge geographical distances and immerse themselves in remote locations as if they were there. This transformative capability promises unprecedented levels of engagement, communication, and investigation across numerous industries, from business and education to healthcare and entertainment [2].

The evolution of sixth-generation (6G) networks is expected to introduce a wide range of telepresence services beyond what is now available. These future networks are expected to have a significant impact, mainly because of their capacity to support various advanced telepresence services [3]. Telepresence services are undergoing a significant transformation driven by the incorporation of holography. The potential of holographic services is anticipated to span a wide range of applications that enable immersive interactions, surpassing physical limitations with exceptional realism [4]. The potential outcomes encompass various domains, including remote collaboration, communication, surgery, education, and entertainment [5].

Telepresence provides many benefits, including cost savings, enhanced productivity, less delay, and high user experience of remote fields. Numerous corporations and platforms offer telepresence services, presenting diverse solutions designed to accommodate various requirements and preferences. These services have experienced a significant rise in popularity, particularly in the era of digital technology, due to their ability to provide effortless communication and collaboration regardless of geographical barriers [6].

Telepresence services' quality of experience (QoE) refers to the set of criteria and parameters that ensure a satisfactory and immersive user experience when participating in remote communication, collaboration, or virtual presence scenarios [7]. These parameters include reliability of communicated data, latency, and network availability. Telepresence services aim to make participants feel as if they are physically present in a remote location, which requires addressing various technical and perceptual factors [5]. This requires providing telepresence services using ultra-reliable low-latency communications (uRLLC) to meet the QoE demands. This requires a one-millisecond delay, full network coverage, and a reliability of 10−5. Such requirements pose many constraints on developing communication systems for delivering telepresence services [8].

The architectural conception of telepresence involves designing the underlying framework and components that enable seamless communication and interaction in virtual environments. It is a critical component in evolving telepresence services and meeting the required QoE of such services [9]. Several models and approaches can be considered for architecting telepresence systems. However, such traditional techniques fail to meet the QoE demands of telepresence services, and thus, novel approaches should be introduced [10].

This chapter considers providing a novel architectural conceptual framework for telepresence services that meet the required QoE. The framework provides the main layers required to deliver the telepresence services. The protocols needed for each layer are introduced. Moreover, we provide use cases of telepresence services that can be implemented using the proposed architecture. The contributions of this chapter are as follows.

1. **Identification of telepresence evolution:** The work identifies telepresence services' main features and characteristics. Also, it identifies the main challenges of implementing telepresence services. The work recognizes the development of communication networks and their impact on remote interaction and collaboration. It emphasizes the transformative effect of telepresence services, which have enabled cross-geographic real-time communication. This includes the 3rd Generation Partnership Project (3GPP) and International Telecommunication Union (ITU) recommendations.
2. **Highlighting the significance of the architectural conceptual model:** The work highlights the critical role of the architectural conceptual model in the evolution of telepresence services. It recognizes that the

design and structure of telepresence systems significantly impact the QoE of telepresence services.

3. **Providing the limitations of traditional approaches:** The chapter investigates traditional architectural conceptual models for providing telepresence services, highlighting their failure to satisfy the QoE requirements for telepresence services. This observation establishes the basis for the requirement of novel and innovative approaches. This encourages us to introduce the proposed work.
4. **Proposing a novel architectural conceptual framework:** The primary contribution of this work is the introduction of a revolutionary architectural conceptual framework designed specifically to provide telepresence services at the required QoE.
5. **Identifying layers architecture:** The proposed framework is characterized by its multi-layered architecture, where each layer is carefully designed to meet the complex requirements of telepresence services. Each layer's architecture, including the required protocols, is defined.
6. **Identifying use cases:** The work identifies six main use cases of the developed conceptual model, including augmented reality (AR), virtual reality (VR), and holographic services. These use cases illustrate how the proposed architecture can be implemented to produce significant telepresence experiences.

8.2 Telepresence Services

Telepresence services refer to technologies and solutions that enable users to have a remote physical presence in a different location. It enables individuals or teams to participate in meetings, events, or collaborations as if they were physically present despite their actual geographic locations. Typically, these services create an immersive and interactive experience by employing high-quality video, audio, and sometimes augmented reality. They enable participants to see, hear, and interact in real-time, creating a sense of presence and a more engaging remote communication experience [11].

There are numerous applications for telepresence services across industries, including business meetings, conferences, education, healthcare, and remote work collaborations. Many vendors and platforms provide telepresence services and solutions, meeting various requirements and preferences [5]. These services have become increasingly popular, particularly in the digital age, enabling seamless communication and collaboration regardless of distance.

Establishing communication networks with uRLLC is one of the most promising trends in the development of future networks and communication systems, e.g., 6G. Initial research into such types of networks began with the Tactile Internet approach [8, 12]. However, since then, the selection of programs that function well with ultra-low latency communication networks has dramatically grown. The round-trip delay metric, which includes values between 1 and 10 ms, is now the fundamental issue for classifying network services within the paradigm of ultra-low latency communication networks [13]. This change represents a development in our understanding of how rapid, seamless, and immediate communication can be done across a wide range of areas.

The fact that this range meets the requirements for AR services with a 5-ms latency is of utmost importance to our inquiry [8]. Due to the expected growth of augmented reality services, this time frame is of great importance in many fields, including healthcare, manufacturing, and instruction. The network traffic that these services generate is also completely new. This introduces new attributes and features that should be considered during communications [12, 13]. For example, when using augmented reality services that provide a full 360° field of view, it is important to take into account the user's angular velocity. This metric indicates the user's interaction speed with the augmented environment and directly impacts the AR user experience [14].

When dealing with the remote manipulation of robotic systems, particularly, latency constraints are critical. In these systems, e.g., the robot-avatar paradigm, the robots mimic human limb movements and are directed by a human or another robot through network interactions. Full tactile experience can be achieved if the end-latency of the communicated data exceeds 1 ms [8, 13].

In accordance with the most recent revision of the 3GPP standard, as detailed in the March 2022 release of TS22.261 [15], specifically Section 7.11 titled "Key performance indicators for haptic and multi-modal services," the communication parameters are specified for the remote control of highly mobile robots capable of achieving up to 50 kilometers per hour speeds. In this context, the acceptable range for the circular delay is between 1 and 20 ms. It is necessary to emphasize that these thresholds are only applied in remote robot control activities conducted within a delimited geographical area of no more than 1 km^2 [15].

It is significant to highlight that the critical need to achieve a latency of 1 ms for the remote control of robots has been theoretically proposed and experimentally verified. This verification has been carried out within the framework of the telepresence services network introduced by the SPbSUT's Department of Communication Networks and Data Transmission [16]. The experimental verification went beyond theoretical investigation by showing that the 1-ms latency standard may be met even over noticeably greater distances.

TS22.26 is a comprehensive framework outlining the requirements for modern services in fifth-generation (5G) networks and beyond, including future networks, e.g., 6G, and the anticipated 2030 networks [15]. Also, the document highlights the aspects regarding remote manipulation of robotic entities. However, the significance of this report goes beyond the context of robotic control. It emerges as a comprehensive repository of guiding principles for dynamic AR and VR services. It also introduces a section on the interaction between VR/AR devices and the associated VR/AR servers. It emphasizes how crucial it is for the VR/AR device and its related server to respond instantly to user input.

TS22.26 provides an outline of needs and a conceptual framework, identifying and addressing the challenges related to telepresence services over the next generation of communication networks. However, it does not provide a full conceptual framework with dedicated protocols for delivering telepresence services [15].

The field of holographic telepresence applications raises significant issues regarding the factors of delay and the inherent temporal delays in their functioning that need to be considered. In this context, an important insight arises from the research conducted in [17]. The authors highlighted the constraints associated with adopting standards for holographic-type communication (HTC) services. The authors introduced audio and video conferencing as the standard basis for HTC evaluation.

This necessitates compatibility with the latency constraints identified for audio and video transmission in conference settings. In particular, the current requirement claims a delay of 320 ms for streaming video, with a strict reliability requirement of 0.99999, indicating an extremely high level of operational robustness [8, 17]. The 3GPP technical specifications [18] included additional audio and video conferencing criteria. Specifically, the specifications include voice streaming requirements with a delay of 280 ms and the existing 320-ms delay requirements for video streaming. Another aspect of critical importance is the search for optimal synchronization between audio and video streams. The allowed time difference between audio and video streams is 40 and 60 ms before and after the video stream, respectively.

The technical parameters of these holographic telepresence services are challenging. According to [17], there is ongoing consideration of the implementation of bandwidth speeds ranging from 2 Gbit/s to an impressive 2 Tbit/s to facilitate the seamless delivery of immersive holographic telepresence experiences. Moreover, the authors of [19] recommended 4.32 Tbit/s speed to meet the QoE of holographic telepresence services. The convergence of Tactile Internet and HTC is expected to result in a rapid emergence of holographic telepresence services. These services are anticipated to become pillars not only in the domain of digital medicine but also in Industry 4.0. Additionally, this combination initiates a paradigm shift, representing a new era with distinct latency and reliable demands.

In [17], the authors investigated telepresence services handling VR, AR, and holographic communications. They considered a use case of controlling distant devices via haptic devices. They achieved distance control using the three different technologies and recorded the user experience in each case. The primary determinant in delivering the service was the user's perception of the distant environment, which changed for each used technology. Table 8.1 summarizes the main differences between the user experiences in each case.

With a hologram, the user has greater control over how closely they examine and interact with distant items [20]. This benefit comes at the expense of requiring extremely low latency and significantly increased throughput, which is especially noticeable when compared to the operating needs of VR and AR applications. Enhanced engagement using holographic technology

Table 8.1 Comparison between user experiences of VR, AR, and holographic systems.

System	Used device	Mechanism	User experience
VR	Virtual reality head-mounted display (VR-HMD)	VR-HMD was utilized to project the remote side with a 360° view	360° videos
AR	Augmented reality head-mounted display (VR-HMD)	VR-HMD was utilized to project volumetric videos of objects on distant sides	Volumetric videos
Hologram	Hologram	A hologram machine was utilized to project the real distant side as a hologram	Holograms

gives an intriguing scenario where latency and data transmission rates must be carefully considered [21].

The ITU and 3GPP approved a preliminary set of telepresence services [19, 22]. Table 8.2 summarizes the main characteristics and requirements of the announced preliminary set of telepresence services, including delay and reliability. The reliability requirements are expressed in terms of the probability of correct packet delivery [8, 15, 18, 22].

The present advancement in conventional communication networks is insufficient to ensure the provision of telepresence services characterized by the QoS and QoE parameters [23]. Standardizing delays within 280 ms for audio communications and 320 ms for video transmissions provides evidence for this claim [18]. For these delay constraints, no special arrangements are required to constrain the span extending from the source of information to the destination when they are in the same geographical region. Indeed, even in speech transmission, where the limit to ensure high QoS is 100 ms, the ITU-T recommends two diagonals of about 5000 km where such quality is guaranteed to be achieved [22]. For Europe, this is the Moscow-Lisbon diagonal; for North America, this is the Seattle-Daytona Beach in Florida.

Table 8.2 Characteristics of the preliminary set of telepresence services.

No.	Telepresence service	Delay (ms)	Reliability	Standard
1	Holographic images	5 ms	0.999999	ITU-T Recommendation
2	AR for medical and industry automation	5 ms	0.999999	ITU-T Recommendation
3	Glove and artificial hand-based robots for performing human economic activities	1 ms	0.999999	3GPP, TS 22.261
4	Manipulator robots and their communities for distant control	1 ms	0.999999	3GPP, TS 22.261
5	Telepresence services for people with disabilities, including the extension of sign language	100 ms	0.9999	ITU-T P.800.1.
6	Teleconferences	• 320 ms for video • 280 ms for audio • Audio can be 40 ms ahead of video, 60 ms behind	0.99999	3GPP TS 22.263

In [22], the authors defined the delay values that arise when using high-speed fiber optics. The estimated delay was 5 Âţs per kilometer, which puts a high constraint on the transmission distance when delivering uRLLC. The preliminary set of telepresence services introduced in Table 8.1 requires an end-to-end delay of 1–5 ms to ensure the QoE. This limits the communication distance to a circle of 50–250 km radius [8, 13]. The distance is considerably decreased considering other delays, e.g., processing delay. Moreover, this delay will be affected by the fiber-optic channels of the core network [13, 24]. Thus, it is critical to investigate population density features, distance between main centers, and gross rating point (GRP) to assess whether the telepresence services could be fully or partially deployed everywhere.

It is crucial to emphasize that the difficulties with uRLLC, including telepresence services, have prompted the development of revolutionary technological approaches that can provide these services within the scope of 5G and next-generation systems, including 6G [25]. Intelligent edge and fog computing, microservices and their associated migration processes, unmanned aerial vehicles, software-defined networking, and the paradigm of device-to-device (D2D) interactions are among the most significant of these revolutionary technologies [26, 27]. Together, these technologies facilitate the way for the development of uRLLC services and the delivery of such services with the required QoE without any constraints on communication distances [12].

8.3 Traditional Conceptual Models for Telepresence Services

Architectural conceptual models play a crucial role in influencing the design of telepresence systems. These models help designers visualize and represent the system's overall structure, components, and interactions before its implementation. The architectural conception of telepresence investigates the complex process of developing the fundamental structure and complex elements that facilitate seamless communication and interaction in virtual environments. These models influence the design of telepresence systems in the following ways [28–30]:

- **System interpretation:** Conceptual models assist in realizing the various system components, including network infrastructure and interfaces. Such models provide a high-level overview, allowing designers to identify and define the telepresence system's essential components.

- **Functional specifications:** By developing conceptual models, designers can determine the functional requirements of telepresence systems, such as the capacity to transmit, support multiple users, and facilitate remote control, among others. These models aid in defining the system's intended functionality and behavior.
- **System integration:** Conceptual models facilitate the identification and integration of various telepresence system components. These models offer a framework for connecting network infrastructure, allowing designers to determine these elements' optimal placement, arrangement, and connectivity.
- **Identifying user experience:** Telepresence systems aim to provide an immersive user experience. Conceptual models facilitate visualization of the user interface, input and output mechanisms, and interaction flows for designers. By considering the user's viewpoint, designers can ensure the system is user-friendly and meets the intended user experience objectives.
- **Providing scalability and flexibility:** Conceptual models meet the scalability and flexibility of designing telepresence systems. These systems allow designers to consider the possibility of future expansion, adaptability, and integration of additional functionalities or features. This ensures that the system can adapt to changing requirements and technological advances.

Traditional models fail to meet the QoE demands of telepresence services. In the following, we provide the limitations of the existing models.

a) **Client-server model:** This model utilizes a centralized server to coordinate data distribution and user communication. Clients, which may be individuals accessing the virtual environment, interact with the server to receive real-time updates and transmit their actions. This model assures synchronized experiences for multiple clients but presents challenges regarding server load and latency [31].

b) **Peer-to-peer model:** In this method, users share data and communicate directly with one another without needing a central server. This can reduce latency and increase network resilience; however, maintaining synchronization and security may be difficult [31].

c) **Distributed architecture:** This model combines the client-server and peer-to-peer models. It employs a network of interconnected elements that manage data and communication collaboratively. This method seeks an equilibrium between synchronization and decentralization [32].

d) **Cloud-based model:** This model offloads computation and storage to remote servers. This can improve scalability and accessibility, allowing more users to participate in telepresence without being limited by local hardware. However, it introduces high latency and low reliability.

Telepresence technologies can be conceptualized as an architectural framework applicable to communication networks. This framework entails a precisely delineated set of flexible tools to facilitate the creation and introduction of new services into the communications network, including services subject to user control. The conceptual model for providing telepresence services establishes a set of rules that do not depend on the service being created and the network structure in which the service will be provided. It is based on three fundamental features that possess the realization and service-providing tools. These features are independent of the service type, the network structure, and the equipment manufacturer.

The independence from the type of services implies that the principle of construction and architecture of the models remain constant, and the changes relate only to the set of functions. In order to ensure this, a specified number of elementary functions, i.e., function blocks, are repeatedly used by the designer to create a variety of services. Independence of functional blocks from the type of services radically facilitates the processes of their specification and creation. In this context, an analogy to the modular construction of buildings emerges. This is similar to an architect who, despite working with a limited number of building materials, is able to design unique homes with distinct characteristics. Independence from the network structure means the ability to flexibly place network functions and resource models and effectively manage them without burdening the features of different equipment based on which the communication network is developed.

8.4 The Proposed Conception Model for Providing Telepresence Services

8.4.1 Model structure

The concept of telepresence service models is based on a "top-down" standardization process, i.e., from the general to the particular, and moving forward as a compromise is reached. The ultimate aim of the proposed conceptual model is to establish a service delivery model architecture that theoretically realizes the proposed concept's three fundamental properties. The conceptual model is proposed as a general framework for all the primary sets of telepresence services, which reflects this architecture on different planes, giving varying degrees of detail.

Figure 8.1 provides the proposed conceptual model for providing telepresence services. The model contains four planes on top of each other, each of which is an abstract representation of a network's capabilities with its degree of detail. It is important to emphasize that this model has no relation with the well-known seven-level model of interaction of open systems, each level of which describes different objects, i.e., protocols.

The proposed conceptual model abstractly represents functions related to one or more services, the possibility and necessity of distributed functioning, as well as the links between the elements of the architecture. The model represents them by functional objects defined on each plane, and the functional objects of neighboring planes relate to each other in a given way. With the help of the model under consideration, it is possible to design services and simulate the process of their provision for networks with different structures and principles of organization.

The upper layer of the model, the service layer, represents services as they are "seen" by the end user. This representation does not contain information related to the ways and details of how the service is implemented in

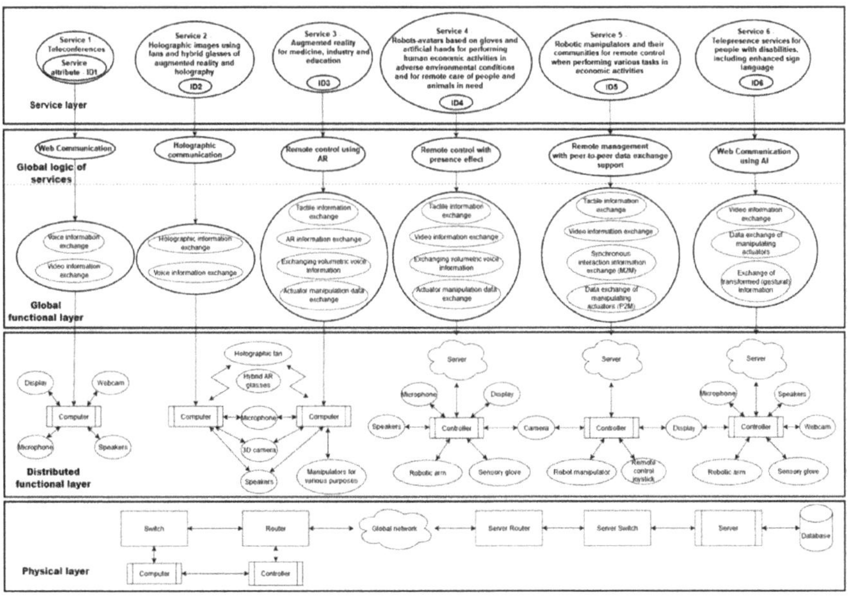

Figure 8.1 The general model for delivering a prioritized set of telepresence services.

the network. This plane shows that services are composed of one or more standardized components, each of which is perceived by the user as one of the characteristic properties or, similarly, as one of the attributes of a service, i.e., service features. A set of such components and the rules for their use are defined for each stage of standardization. The global functional layer contains a set of independent services functional blocks, presented as a single constructive object.

The distributed functional layer shows that the implementation of the service in the environment of the concept under consideration is performed by software in a distributed manner. Each functional object defined on this layer can perform several actions defined for it. The independent considered building blocks represented on the distributed functional layer as a model that combines functional components and displays information flows between objects.

The physical plane represents the physical entities of the network in which the telepresence service concept is implemented. These physical entities can be switching stations, specialized computers, or databases. The physical plane also shows which protocols support the information exchange between the different physical network entities.

The logic of the processes that provide telepresence services is represented differently on different layers. On the global functional layer, only the global service logic is visible, which contains service modules, each of which supports one of the service components, i.e., one of the attributes. The same logic is seen as a set of distributed functional components on the distributed functional layer. Finally, the physical layer shows that the network devices that implement the service logic are placed in the network's designated hardware and software nodes.

8.4.2 Connections between layers

The concept of telepresence service provision specifies a strictly defined mutual correspondence between the architectural elements of the model layers. The service attributes introduced on the service layer are implemented on the global functional layer by appropriate combinations of functional blocks and service modules of the global service logic.

Each functional unit introduced on the global functional layer must be represented in at least one functional object of the distributed functional layer. Alternatively, it can also be implemented in several functional objects. In the last case, joint actions of different objects may be required. Connections between different functional objects, visible on the distributed functional layer, are specified on the physical layer as protocols.

8.4.3 Architecture of layers

8.4.3.1 Architecture of the service layer

A service, as applied to the concept of telepresence, is defined as a separately offered product formed from one or more mandatory components and can be modified by including other components. Figure 8.2 presents a primary set of telepresence services deployed in the service layer. A service component is a product offered either as a mandatory part of one or more services or as an additional part that allows to modify the service(s). In both cases, each component on the service layer represents one of the properties, i.e., attributes, of the service it belongs to. This allows the network operator to demonstrate how the network can provide one or another service to a potential client.

The telepresence network operator offers the potential customer a list of service attributes for each provided service. The customer chooses the desired

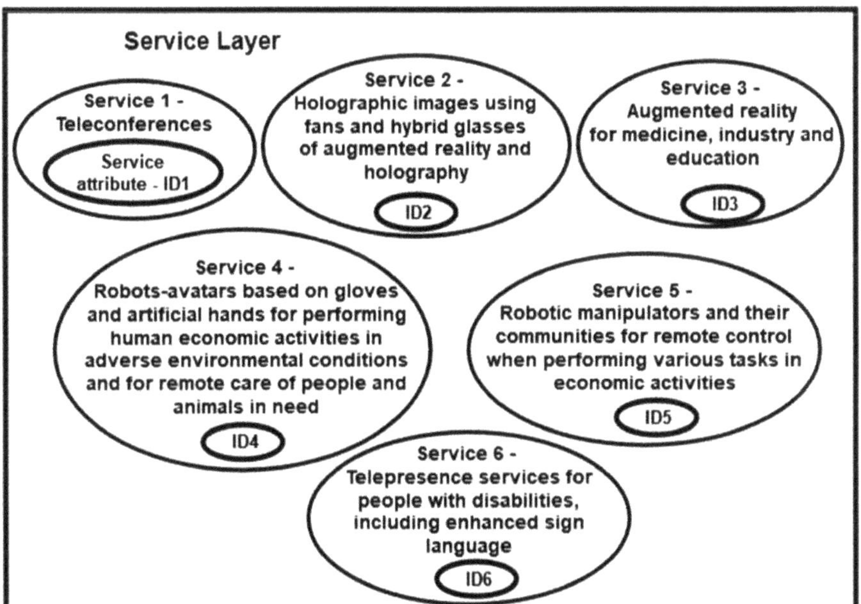

Figure 8.2 A primary set of telepresence services.

features and describes how he envisions the service having these attributes. The result would be a specification of a service for a particular customer.

In the next step, the designer models this service on each layer of the conceptual model. Modeling on the service layer is relatively trivial and comes down to using service attributes as layout modules for its specification and creation. The critical point is that the service component, which forms an attribute, represents the smallest inseparable part, which can be demanded by the user when designing the service to meet his requirements.

Using the capabilities of telepresence service models, it is much easier and faster for an operator to provide a service even to a narrow group of clients for a limited time. This meets the market requirements of modern telecom services regarding the ability to work with a large number of relatively small groups of users, each of which has its own dynamic requirements.

8.4.3.2 Architecture of the global functional layer

The functions provided on this layer are not specific to concrete services or their constituents. These functions are referred to as service-independent functional blocks. Figure 8.3 provides the architecture of the global functional layer. For a service represented on the service layer as a set of components, the global functional layer defines a correspondence between each component and one or more functional units. By its essence and definition, no functional block depends on the service and, therefore, cannot contain information about the subsequent development of events. The only element on the global functional layer that depends on services and includes information on the sequence of block execution is the global logic of services.

Global logic stores the information necessary to form a chain of functional blocks, information about the sequence of their connection and solution options, and the data required for each. The main feature of the global functional layer is the possibility of multiple uses of functional blocks during the construction of new services. Detailed implementation of functional blocks on this layer is invisible, and they appear as indivisible components with a defined interface. However, the significance of designing services using the concept under consideration is that this fact is hidden from the creator of a new service. Functional blocks are implemented by actions of one or more functional objects of the distributed functional layer.

A finite chain of functional blocks can describe any service component. If the blocks do not depend on services, then the creation of a chain, which forms one or another component of a service, each chain link acquires some elements of dependence on this component. This dependence is expressed

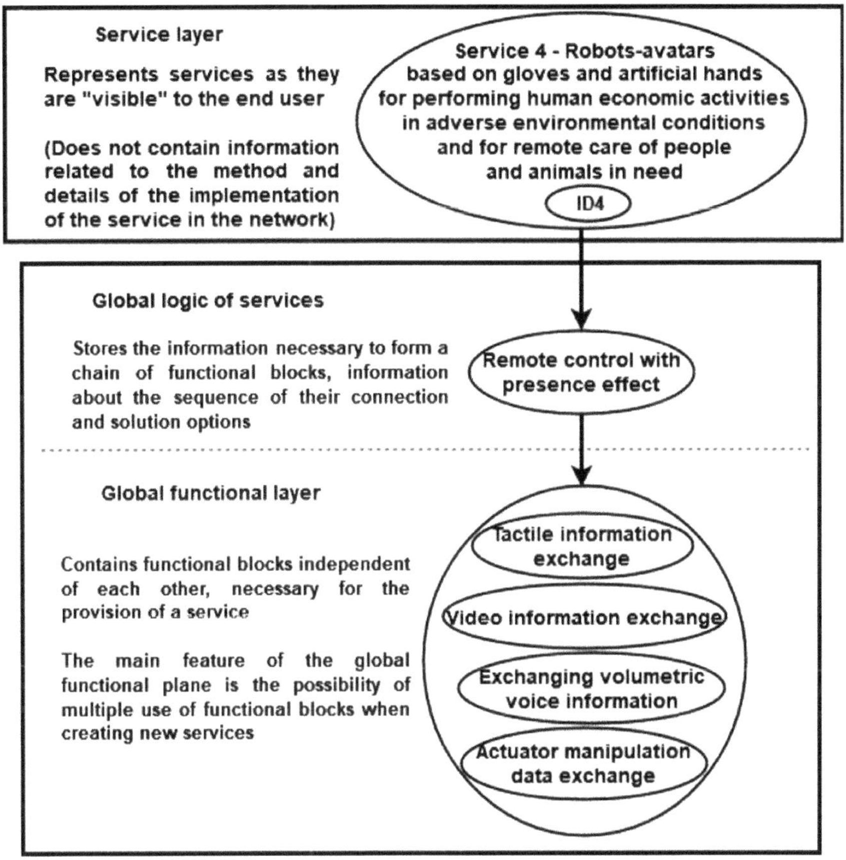

Figure 8.3 The architecture of the global functional layer.

through the parameters specified for each functional block and set by the global logic of services. Global logic defines the order in which functional blocks must be connected to be executed consistently in provisioning a particular service. This logic can be likened to glue that connects building blocks in the right order.

8.4.3.3 Distributed functional layer architecture

The distributed functional layer provides the principles of interaction and information flows, i.e., connections between functional objects arising in service provision. Figure 8.4 presents the architecture of the distributed functional layer, which has a standardized design concept for all sets of

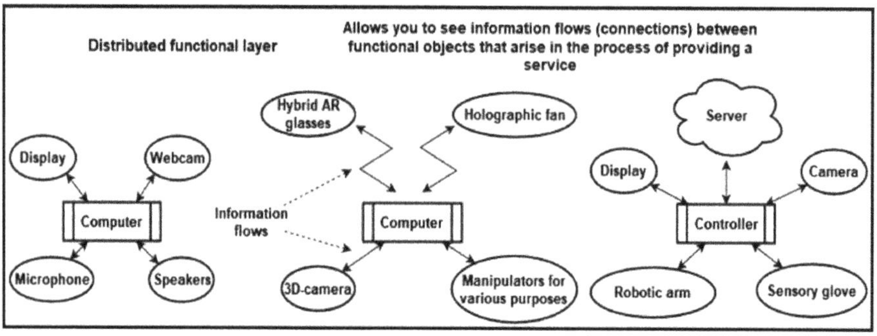

Figure 8.4 The architecture of the distributed functional layer.

telepresence services. The ovals show the functional objects placed on this plane, each representing a unique group of functions concentrated in a single physical element and a subset of the complete set of functions required to implement the service.

The interaction between functional objects in the model is called the information flow. The connection between any two functional objects is displayed in the model by a group of information flows denoted by lines connecting these functional objects. A link may have a type identifier that uniquely identifies a group of information that flows in the model. Moreover, several links of the same type may exist in the model.

8.4.3.4 The architecture of the physical layer

The physical layer represents the network's physical elements, interfaces, and protocols. All objects of the distributed functional layer must be mapped to physical layer elements, and one physical element can correspond to one

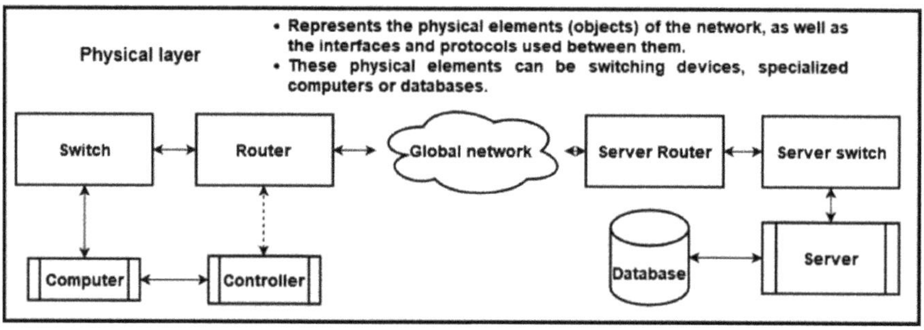

Figure 8.5 The architecture of the physical layer.

or more functional objects. A standard interface and information transfer protocol is defined for the physical elements. Figure 8.5 presents the main architecture of the physical layer.

An element of the physical layer, which contains one or more objects of the distributed functional layer, is considered a network node in the proposed conceptual model of telepresence service provision. Figure 8.1 presents the general structure of the proposed conceptual model of telepresence service.

8.5 Use Cases

Many telepresence services can be implemented over the proposed architectural model. This section considers common use cases that can be implemented using the proposed conceptual model [25].

- **First service – teleconferences:**
  The main attribute of the service is web-communication. Based on the information stored in this attribute, the chain of functional blocks on the global functional layer is formed, and the information about the sequence of their connection and options for placement is formulated. Teleconferencing services should include functional blocks responsible for exchanging voice and video information. Thus, such functional objects, each representing a unique group of functions concentrated in one physical element, should be used on a distributed functional layer as a monitor, a webcam, a microphone, and speakers. Then, a physical layer is activated, which is responsible for the information exchange between the computer and devices outside the local network segment.
- **Second service – holographic imaging:**
  The main attribute of the service is holographic communication. Accordingly, the service must include functional units for exchanging holographic and voice information. Thus, on the distributed functional layer, it is necessary to use the following functional objects:

  a. a holographic fan and hybrid AR glasses reproduce the three-dimensional image on the receiving and transmitting sides involved in the communication,
  b. 3D camera responsible for the formation of three-dimensional images, and
  c. microphone and speakers to capture and playback the voice data stream.

- **Third service – AR for medical, industry automation, and educational services:**
The main attribute of the service is the remote control using AR. Thus, the service should include functional blocks for exchanging tactile, AR, and voice information. Moreover, the functional blocks should support the ability to exchange actuator manipulation data received from sensors. Therefore, the following functional objects should be used on the distributed functional layer:
    a. a holographic fan and hybrid AR glasses reproduce the three-dimensional image on the receiving and transmitting sides involved in the communication,
    b. 3D camera responsible for the formation of three-dimensional images,
    c. microphone and speakers for capturing and playing back the voice data stream, and
    d. manipulators of different purposes that are responsible for physical movement in space, e.g., the robotic hand, crane-loader, and robot-avatar.
- **Fourth service – robotic avatars (based on gloves and artificial hands to perform economic activities):**
The main attribute of the service is the remote control with the effect of presence. Accordingly, the service under consideration should include the functional blocks responsible for exchanging tactile, voice, and video information and support the ability to exchange actuator manipulation data received from sensors. Thus, on the distributed functional plane, it is necessary to use the following functional objects connected to the controller:
    a. camera, which is responsible for generating video images,
    b. monitor for reproducing the image on the receiving and transmitting sides involved in the communication,
    c. microphone and speakers for capturing and playing back the voice stream,
    d. a sensor glove and a robotic hand that are responsible for receiving information about the movements of the controlling person's fingers and the physical movement of the actuator in space, respectively, and
    e. a server, which is involved in transmitting, processing, and storing tactile information.

- **Fifth service – robotic manipulators and communities for remote economic activities:**
  The main attribute of the service is the remote control with the support of peer-to-peer data exchange. The service under consideration includes functional blocks responsible for exchanging tactile and video information and supporting the ability to exchange synchronous interaction information, i.e., machine-to-machine (M2M), and actuator manipulation data, i.e., point-to-multipoint (P2M), received from sensors. Thus, on the distributed functional layer, it is necessary to use the following functional objects connected to the controller:
  a. a camera responsible for generating video images on the remote side,
  b. monitor for reproducing the image on the side controlling the manipulating device,
  c. joystick remote control and robotic hand that are responsible for information about the movements of the arm of the controlling person and the physical movement of the manipulating device in space, and
  d. a server, which is involved in the transmission, processing, and storage of information about coordinate movements.
- **Sixth service – telepresence services for people with disabilities, including sign language extensions:**
  The main attribute of the service is web-communication using artificial intelligence. Accordingly, the service under consideration includes the functional units responsible for video information exchange and support the exchange of transformed information and actuator manipulation data received from sensors. Thus, on the distributed functional layer, it is necessary to use the following functional objects connected to the controller:
  a. webcam, which is responsible for generating video images,
  b. monitor, for playing back the image on the receiving and transmitting sides involved in the communication,
  c. microphone and speakers for capturing and playing back the voice (converted from gesture information) data stream,
  d. sensor gloves and robotic hands, which are responsible for obtaining information about the movements of the hand of the controlling person and the physical movement of the actuator in space

(reproduction in sign language of the information converted from the voice stream), and

e. AI-assisted server, which is involved in the transmission, processing, and storage of gesture information.

8.6 Summary

This chapter provided an extensive investigation of the essential characteristics and fundamental characteristics of telepresence services. Also, it introduced the main challenges of implementing telepresence services. We demonstrated the critical role of architectural conceptual models in the evolution of telepresence services and their significant impact on the QoE of telepresence services. The limitations of the traditional conceptual approaches were introduced. Then, the chapter presented a novel architectural model for handling telepresence services. The main layers, functions, and required protocols for each layer were introduced. We provided common telepresence services that can be implemented over the proposed model. The main attributes and the required steps for handling such use cases were provided.

References

[1] N. Arora *et al.*, "Augmenting immersive telepresence experience with a virtual body," *IEEE Trans. Vis. Comput. Graph.*, vol. 28, no. 5, pp. 2135–2145, 2022.

[2] K. Youssef, S. Said, S. Al Kork, and T. Beyrouthy, "Telepresence in the recent literature with a focus on robotic platforms, applications and challenges," *Robotics*, vol. 12, no. 4, p. 111, 2023.

[3] J. Erfanian, D. Lister, Q. Zhao, G. Wikström, and Y. Chen, "6G vision & analysis of potential use cases," *IEEE Commun. Mag.*, vol. 61, no. 4, pp. 12–14, 2023.

[4] Y. Huang, Y. Zhu, X. Qiao, X. Su, S. Dustdar, and P. Zhang, "Toward holographic video communications: A promising AI-driven solution," *IEEE Commun. Mag.*, vol. 60, no. 11, pp. 82–88, 2022.

[5] A. E. Koucheryavy *et al.*, "A first-priority set of telepresence services and a model network for research and education," in *Communications in Computer and Information Science*, Cham: Springer Nature Switzerland, 2023, pp. 208–219.

[6] A. Tella and O. P. Ogbonna, "Telepresence robots in libraries: applications and challenges," *Libr. Hi Tech News*, 2023.

[7] G. Kougioumtzidis, V. Poulkov, Z. Zaharis, and P. Lazaridis, "QoE assessment aspects for virtual reality and holographic telepresence applications," in *Lecture Notes of the Institute for Computer Sciences, Social Informatics and Telecommunications Engineering*, Cham: Springer International Publishing, 2022, pp. 171–180.

[8] A. A. Ateya, A. Muthanna, M. Makolkina, and A. Koucheryavy, "Study of 5G services standardization: Specifications and requirements," in *2018 10th International Congress on Ultra Modern Telecommunications and Control Systems and Workshops (ICUMT)*, 2018, pp. 1–6.

[9] K. Tonchev, I. Bozhilov, R. Petkova, V. Poulkov, A. Manolova, and P. Lindgren, "Implementation requirements and system architecture for mixed reality telepresence application scenario," in *2021 24th International Symposium on Wireless Personal Multimedia Communications (WPMC)*, 2021, pp. 1–6.

[10] A. Yastrebova, R. Kirichek, Y. Koucheryavy, A. Borodin, and A. Koucheryavy, "Future Networks 2030: Architecture & Requirements," in *2018 10th International Congress on Ultra Modern Telecommunications and Control Systems and Workshops (ICUMT)*, 2018, pp. 1–8.

[11] Z. Ying, D. Ghadiyaram, and A. Bovik, "Telepresence Video Quality Assessment," in *Lecture Notes in Computer Science*, Cham: Springer Nature Switzerland, 2022, pp. 327–347.

[12] M. Osama, A. A. Ateya, S. Ahmed Elsaid, and A. Muthanna, "Ultra-reliable low-latency communications: Unmanned aerial vehicles assisted systems," *Information (Basel)*, vol. 13, no. 9, p. 430, 2022.

[13] A. A. Ateya *et al.*, "Model mediation to overcome light limitations—toward a secure tactile Internet system," *J. Sens. Actuator Netw.*, vol. 8, no. 1, p. 6, 2019.

[14] A. A. Ateya, A. Muthanna, A. Koucheryavy, Y. Maleh, and A. A. A. El-Latif, "Energy efficient offloading scheme for MEC-based augmented reality system," *Cluster Comput.*, vol. 26, no. 1, pp. 789–806, 2023.

[15] 3GPP, "TS 22.261. Service Requirements for the 5G System," Stage 1 (Release 18), v.18.6.1, Jun. 2022.

[16] "MEGANETLAB 6G." [Online]. Available: https://www.sut.ru/meganetlab. [Accessed: 15-Aug-2023].

[17] T. Taleb, Z. Nadir, H. Flinck, and J. S. Song, "Extremely-Interactive and Low Latency Services in 5G and Beyond Mobile Systems," *IEEE Communications Standard Magazine*, vol. 5, pp. 114–119, 2021.

[18] 3GPP, "TS 22.263. Service requirements for video, imaging and audio for professional applications (VIAPA)," Stage 1 (Release 17), Jun. 2021.

[19] M. T. Vega, "Towards Truly Immersive Holographic-Type Communication: Challenges and Solutions," in *4th ITU Workshop on Network 2030. S.-Petersburg, Russia*, 2020.

[20] H. Zhang *et al.*, "Holographic integrated sensing and communication," *IEEE J. Sel. Areas Commun.*, vol. 40, no. 7, pp. 2114–2130, 2022.

[21] E. Wolf, M. L. Fiedler, N. Dollinger, C. Wienrich, and M. E. Latoschik, "Exploring presence, avatar embodiment, and body perception with a holographic augmented reality mirror," in *2022 IEEE Conference on Virtual Reality and 3D User Interfaces (VR)*, 2022, pp. 350–359.

[22] Recomendation, "Y.154. Network performance objectives for IP-based services. Telecommunication Standardization Sector of ITU," ITU, Geneva, 2018.

[23] M. Conde, V. Mikhailova, and N. Döring, "Towards augmented reality-based and social robot-based social integration of older adults: A user requirements analysis," in *Communications in Computer and Information Science*, Cham: Springer Nature Switzerland, 2023, pp. 426–432.

[24] C. Ranaweera *et al.*, "Design and deployment of optical x-haul for 5G, 6G, and beyond: progress and challenges [Invited]," *J. Opt. Commun. Netw.*, vol. 15, no. 9, p. D56, 2023.

[25] R. A. Dunaytsev *et al.*, "A First-Priority Set of Telepresence Services and a Model Network for Research and Education," in *Distributed Computer and Communication Networks: 25th International Conference*, Moscow, Russia: Springer Nature, 2022.

[26] A. Muthanna, A. A. Ateya, M. A. Balushi, and R. Kirichek, "D2D enabled communication system structure based on software defined networking for 5G network," in *2018 International Symposium on Consumer Technologies (ISCT)*, 2018, pp. 41–44.

[27] M. Shahjalal *et al.*, "Enabling technologies for AI empowered 6G massive radio access networks," *ICT Express*, vol. 9, no. 3, pp. 341–355, 2023.

[28] J. Guan and A. Morris, "Design frameworks for hyper-connected social XRI immersive metaverse environments," *arXiv [cs.HC]*, 2023.

[29] D. G. S. Pivoto *et al.*, "A detailed relevance analysis of enabling technologies for 6G architectures," *IEEE Access*, pp. 1–1, 2023.

[30] K. Lee, J. Yi, and Y. Lee, "FarfetchFusion: Towards Fully Mobile Live 3D Telepresence Platform," in *Proceedings of the 29th Annual International Conference on Mobile Computing and Networking*, 2023.

[31] L. Wu, W. Lu, and C. Chen, "Strengths and weaknesses of client-server and peer-to-peer network models in construction projects," *Int. J. Constr. Manag.*, pp. 1–15, 2023.

[32] W. Xu, Z. Yang, D. W. K. Ng, M. Levorato, Y. C. Eldar, and M. Debbah, "Edge learning for B5G networks with distributed signal processing: Semantic communication, edge computing, and wireless sensing," *IEEE J. Sel. Top. Signal Process.*, vol. 17, no. 1, pp. 9–39, 2023.

9

Methods for Testing Telepresence Networks

9.1 Introduction

Communication network testing techniques are ways to test the functionality and effectiveness of a network. They are carried out both when deploying new networks and when servicing existing ones. Testing techniques aim to identify possible problems in the network and help optimize its operation. Depending on the purpose of testing, different techniques are used. Some of them are aimed at checking the data transfer speed, others at checking the stability of the network, its security or compliance with standards. In any case, testing allows you to identify shortcomings and errors in the network, which helps improve the quality of its work. There are various tools and programs that are used in testing. For example, to check the stability of the network, programs that create a large amount of test data can be used, and to check security, special laboratory benches can be used on which simulations of attacks and defensive actions are carried out. It is important to consider that testing methods may differ depending on the type of communication network, its topology, and purpose. Therefore, when choosing a technique, you should consider the characteristics of a particular network and its needs.

The scope of testing of communication networks includes the following:

- Checking the possibility of the functioning of all three system parts simultaneously based on the requirements and technical specifications described.
- Assessing the performance of modules of ultra-dense network nodes, traffic generators, and receivers and possibly constructing a network using Hilbert curves and other fractal constructions based on the requirements and technical specifications described. At this stage, the performance of the ultra-dense network traffic monitoring device is also assessed.

- Checking the functionality of the augmented reality and holographic images subsystem, including preparing a stand for transmitting images to a projector, transferring images to the projector at different rates of communication network bandwidth and subjective assessment of their quality, as well as preparing a stand for transmitting images to a holographic fan, transferring images to holographic fan and subjective assessment of their quality based on the requirements and technical specifications described.
- Checking the performance of the subsystem of avatar robots and multifunctional robots based on the requirements and technical specifications described. This stage also includes setting up the control equipment for avatar robots, assessing the influence of channel attenuation on the circular delay of the transmission of control data for avatar robots, setting up the equipment of multifunctional robotic manipulators, and assessing the influence of channel attenuation on the transmission of control data for multifunctional robotic manipulators.

Several methods for testing communication networks allow you to identify possible problems in the network and optimize its operation. This includes the following.

1) **Performance testing:** This method allows you to determine what data transfer speed the network can provide in real conditions. For this purpose, special programs are used that simulate the transfer of a large amount of test data.
2) **Stability testing:** This technique checks how reliably the network operates over time and how quickly it recovers after failures. To do this, programs are used that generate data streams and load the network.
3) **Security testing:** A technique for checking how securely the network operates and how protected it is from hacking and attacks. For this purpose, laboratory benches are used that allow you to simulate attacks and ways to prevent them.
4) **Compliance testing:** This method allows you to evaluate the network's compliance with standards and requirements established in the telecommunications industry. For this purpose, special programs and tools are used compared to the prescribed requirements. However, not all methods apply to every communication network, and specific testing techniques may be necessary. It is also important to consider the characteristics of the network, its topology, and its purpose in choosing the most effective testing method.

Each of these techniques has its own strengths and weaknesses; so it is recommended to use them in combination for the most complete testing of communication networks. Using these methods, responsible organizations and government agencies can ensure a high level of security for their communications networks.

Additionally, there are a number of tools that help automate communications network testing, making it more efficient and cost-effective. These tools are often used in conjunction with testing techniques to make testing easier and faster.

Modern technologies such as the IoT and cloud computing also pose new challenges for testing communications networks. Testing IoT devices requires determining how they interact with other devices on the network, as well as identifying vulnerabilities that could be exploited to hack them. Testing cloud communication networks requires determining the level of security of cloud resources and carefully checking the configuration and settings.

9.2 Challenges and Limitations

Despite the fact that current methods for testing communication networks are constantly being improved, they also have their limitations. The first limitation is the difficulty of testing the entire network as a whole. In today's multi-device communications networks, it is difficult to test each component individually to ensure it is functioning properly in the context of the entire network. The second limitation is that it is not always possible to conduct testing under realistic conditions. For example, when testing under heavy load, there may be problems with the availability of the network under test, which can complicate the testing procedure.

The third limitation is the difficulty of identifying all possible vulnerabilities in network security. Security testing may be limited to identifying only the most common threats, while more complex and unusual vulnerabilities may go undetected. The fourth limitation is the high cost of testing and the need to use special equipment and software. This can mean that many organizations may not have access to meaningful testing of their networks. Thus, current methods of testing communication networks have their limitations, which can significantly complicate the process of identifying and eliminating network problems.

Despite the limitations, modern methods of testing communication networks have many positive things:

1) **Automation of testing:** Modern methods of testing communication networks provide the ability to automate testing, which significantly reduces the cost of time and resources.
2) **Wide range of testing:** Modern testing methods allow you to test all network components, which helps identify various problems related to network operation.
3) **Increased testing speed:** The use of modern testing techniques allows you to test the network and its components at high speed, which significantly reduces the time of interruption in network operation.
4) **Improving the quality of network operation:** Modern testing methods help improve the quality of network operation and eliminate problems associated with its operation, which helps to increase the efficiency of business processes running on the network.
5) **Improving network security:** Modern security testing methods allow you to identify network vulnerabilities and prevent possible attacks on it.

Thus, modern methods of testing communication networks have a number of advantages that help optimize operation and increase network efficiency. Choosing the right testing methodology and applying it optimally through a combination of automated processes and manual checks can help companies optimize their operations and increase profits.

The requirements for testing techniques from communication networks and holographic images are high, as they are a critical component for the operation of modern technologies. For communication networks, requirements include network infrastructure testing, throughput optimization, fault tolerance testing, signal quality testing, and security testing.

Testing also includes performance and data transfer checks, 24/7 network monitoring, and real-time diagnostics of all nodes and connections. For holographic images, requirements include checking image quality and clarity, checking image depth and accuracy, testing for distortion and distortion, and checking color rendition.

Testing holographic images requires specialized equipment such as holographic projectors and cameras, which can result in additional costs and operational complexity. It should also be taken into account that holographic images can often place increased demands on network bandwidth, which should also be taken into account during testing.

In general, testing techniques must meet the high demands placed on modern communications networks and holographic imaging technologies.

This helps ensure the reliability, security, and efficiency of the network and technology that is critical to today's organizations.

9.3 Performance Testing

The performance testing method is one of the most common methods for testing communication networks. This method allows you to evaluate the speed of data transfer on the network and identify possible problems in operation. To test network performance, special programs are used that generate a large amount of test data and transmit it through the network. Programs can check the data transfer speed and the level of network delays (latency) and packet loss (packet loss).

One such program is Iperf [1], available on various operating systems, including Windows, Linux, and MacOS. It allows you to create test data streams with different speeds and directions of data transmission and check the level of delays and packet losses in the network. For example, when using Iperf, you can create server and client parts that will be used to transfer data. You can then determine the desired data transfer rate and run the test. Test results can be presented in the form of graphs, tables, and diagrams, which allow you to analyze network performance and identify possible problems with its operation.

Iperf has two operating modes: server and client. In the server mode, the program waits for connection requests from the client, and in the client mode, the program sends a predetermined amount of data to the server and measures the transfer speed. The main features of Iperf are as follows:

1) Determination of data transfer speed and communication network capacity.
2) Determination of delay (transmission delay) and loss of data packets.
3) Supports TCP and UDP testing modes.
4) Ability to set the number of threads and packet size.
5) Possibility of specifying testing time.
6) Ability to save test results in CSV files.
7) Cross-platform support.

Iperf allows you to determine the performance of not only your local network but also your internet connection. To do this, you need to run the server part of the program on a remote machine and connect to it using the client part. Iperf supports several additional parameters that allow you to manage test traffic and customize the program's behavior. Iperf can also be used in scripts, which allows you to automate the testing process.

Another example of a program that allows you to test a communication network for performance is the JPerf program [2]. It is a free, portable program used to test network performance. It is based on an earlier version of Iperf and is designed to measure throughput, data rate, latency, packet loss, and other network parameters. JPerf provides a simple and intuitive user interface. It allows you to select the testing mode (TCP or UDP), set the number of threads and data packet size, specify the operating time, etc. The program allows you to save test results into CSV files for more detailed analysis. JPerf is a cross-platform; so it can be used on any operating system that supports Java.

JPerf can be used to test networks of any complexity, including local networks and broadband internet connections. It can be used to test network performance at different stages of the network life cycle, including before go-live, during operation, and after changes have been made to the network. JPerf is open source and can be downloaded from the original project's website. However, if you do not want to go through the trouble of installing this software, many internet service providers provide their own network speed testing programs. Figure 9.1 presents the JPerf program interface.

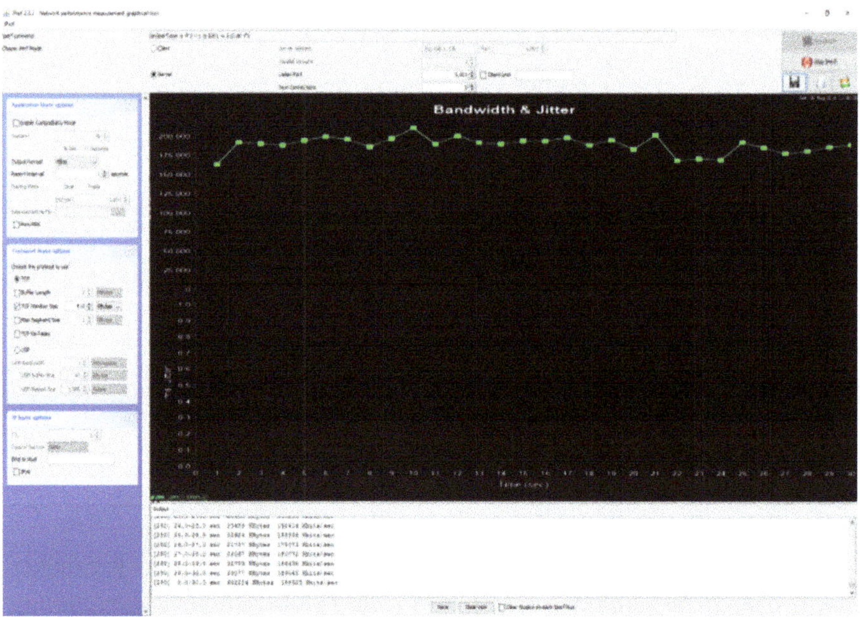

Figure 9.1 JPerf program interface.

9.4 Stability Testing

An example of such a program is the PingPlotter utility [3]. This program monitors the stability and quality of a communication network. It allows you to track problems with the functioning of the network, determine the causes of errors and the duration of downtime, and build reports and graphs based on the test results. PingPlotter uses ICMP technology to send pings to devices on a network to provide feedback on network status and load. It also uses TraceRoute technology to track the route of data packets across the internet and verify that the network has been configured thoroughly. PingPlotter provides many options for testing network stability, including the following:

1) Identifying network problems, such as packet loss, delays in data transfer, and other network load conditions.
2) Identifying problems with network quality, such as connection breaks, delays in data transmission, and other factors affecting network stability.
3) Perform testing in real time with the ability to view graphs and reports. Determine the route of data transmission on the internet and detect routing problems.
4) Providing detailed reports and the ability to set up automatic notifications about network problems.

PingPlotter is a cross-platform program and can be used on any operating system, including Windows and MacOS. The program uses ICMP technology to send pings to devices on the network to receive a response about network status and load. It also uses TraceRoute technology to track the route of data packets across the internet and verify that the network has been configured thoroughly. PingPlotter provides many testing options. Figure 9.2 presents the PingPlotter program interface.

This utility has a simple and intuitive interface. It provides a graphical display of the received information in the form of charts and graphs, which allows you to assess network problems more clearly. Due to its flexibility, PingPlotter can test networks of any complexity, including local networks and broadband internet connections. Overall, PingPlotter is a powerful tool that lets you quickly detect and solve network problems, which is very important for business and everyday computer use.

Another program that allows you to test the stability of a communication network is NetCrunch [4]. It is a network monitoring program that offers many options for diagnosing and identifying problems with network equipment and connections. NetCrunch uses a suite of tools to collect network

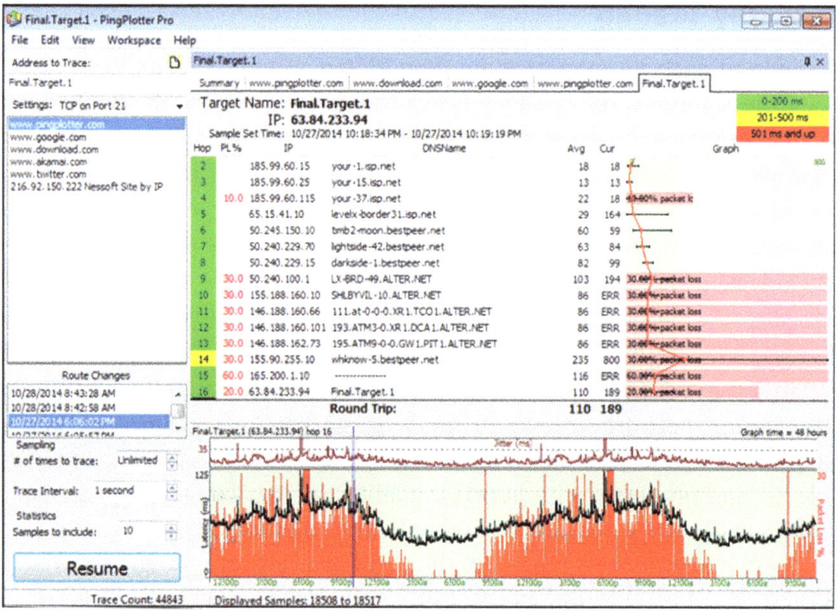

Figure 9.2 PingPlotter program interface.

health data and identify problems using a suite of advanced algorithms and analysis engines. It allows you to monitor your network in real-time and make changes to prevent possible problems. The program uses simple network management protocol (SNMP) technology to collect information about devices on the network, and it can also use ICMP, ARP, WMI, and others to obtain information about the operation of the network. The benefits of NetCrunch include the following:

1) Identifying network problems, such as packet loss, delays in data transfer and other network load conditions.
2) Identifying problems with network quality, such as connection breaks, delays in data transmission and other factors affecting network stability.
3) Providing detailed reports and the ability to set up automatic notifications about network problems.
4) Detection and collection of information from almost any network equipment or service used on the network.
5) Comprehensive monitoring of load, condition, performance, and quality of network operation.

NetCrunch is a cross-platform program that runs on Windows, Linux, and MacOS. It supports a large number of protocols and formats, which allows it

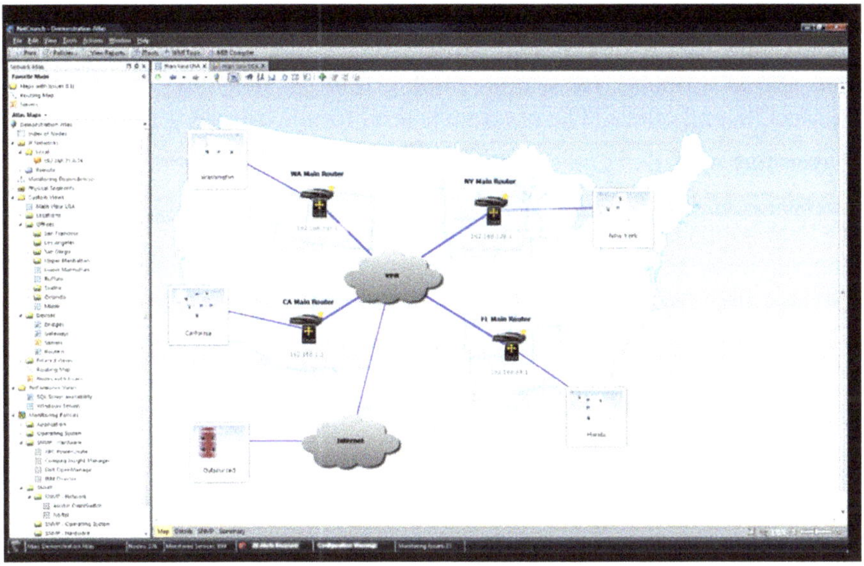

Figure 9.3 NetCrunch program interface.

to work with any network, including local networks and connections over the internet. Figure 9.3 presents the NetCrunch program interface.

9.5 Security Testing

Nmap (Network Mapper) [5] is one of the most popular utilities for exploring a network and checking its security. The program's main functions are to determine the network topology, identify devices, scan open ports, determine the services that run on these ports, and check systems for vulnerabilities. One of Nmap's most important features is scanning open ports. It uses several technologies to determine open ports on devices on a network. This can help you find out what services are running on devices on your network and, therefore, what threats might be associated with those services if they are vulnerable.

With Nmap, you can scan individual devices and entire networks. The program can scan many systems and even use scripts to auto-detect systems on the network. Nmap has a clean and intuitive interface that allows users to easily customize scanning to suit their needs. The program has many options and settings that allow the user maximum flexibility in working with it. However, it is worth noting that you can only use Nmap if everything is correct and legal. Using Nmap to attack networks and computers without the owner's permission may result in criminal prosecution. Nmap is an open-source program and is available for use on most platforms, including Windows, Linux, MacOS, and others. The program is used for the following purposes:

1) Determine open ports on computers or devices on the network.
2) Scanning devices for vulnerabilities.
3) Determination of the network topology and devices in it.
4) Identification of operating systems that are used on computers and devices on the network.
5) Definition of services and applications that run on servers and workstations on the network.

The main advantages of Nmap include the following.

1) High scanning speed.
2) A large number of options and flags for setting up and controlling scanning.
3) Graphical interface support, which makes working with the program easier for inexperienced users.
4) Options that allow you to hide scanning from end users so as not to arouse suspicion or attract the attention of network defenders.

Another program that allows you to test the security of a communication network is Burp Suite [6]. Burp Suite is a web application testing tool that can be used to test network security. It allows you to test all aspects of web applications, including data input and output, request and response processing, authentication, authorization, information leaks, and other vulnerabilities. Burp Suite consists of several modules, including:

1) Proxy is a module that allows you to intercept and modify traffic between a web browser and a server. This allows you to test for data injection vulnerabilities, cross-site code execution (XSS), SQL injection, and other vulnerabilities.
2) Scanner is a module that automatically scans a web application for vulnerabilities. It can be used to scan for XSS vulnerabilities, SQL injections, information leaks, and other vulnerabilities.
3) Intruder is a module that can be used to test authentication and authorization vulnerabilities. It allows you to automate password brute force attacks and other attack methods.
4) Repeater is a module that allows you to repeat requests to the server and change the data sent to the server. This allows you to test vulnerabilities

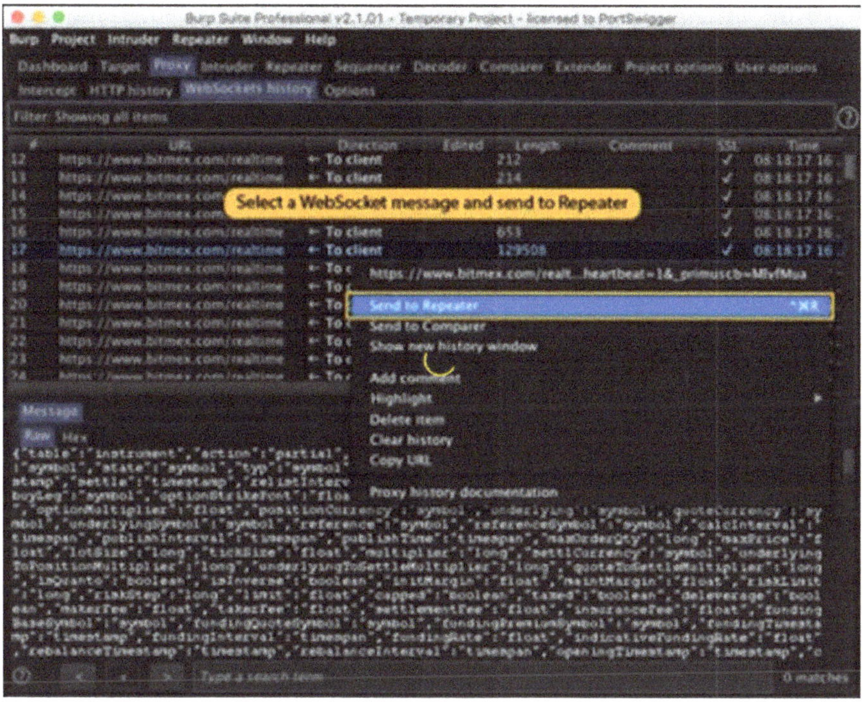

Figure 9.4 Burp Suite program interface.

associated with data input to the server. These modules allow you to test various web application vulnerabilities and use this knowledge to ensure the security of your communication network.

Burp Suite has free and paid versions, but it is recommended to use the latter for full use. It supports Windows, Linux, and MacOS and is available on the official website. Figure 9.4 presents the Burp Suite program interface.

9.6 Compliance Testing

An example of a program that allows testing for compliance with standards is OWASP ZAP (Zed Attack Proxy) [7]. OWASP ZAP is a web application testing tool that allows you to verify compliance with web application security standards. It includes capabilities for vulnerability detection, security compliance testing, performance testing, and compliance testing. Some of the features and functionality of OWASP ZAP include:

1) **Traffic interception:** OWASP ZAP can intercept and trap traffic between the browser and the web application server. It can also analyze traffic content for vulnerabilities such as information leakage, XSS (cross-site scripting), and SQL injection.
2) **Web application scanning:** OWASP ZAP can scan web applications to detect security vulnerabilities. It uses multiple sources of vulnerability information, such as vulnerability databases and attack pattern libraries.
3) **Security compliance testing:** OWASP ZAP can test web applications against certain security standards such as OWASP Top 10, PCI DSS, HIPAA, and many others.
4) **Ancillary tools:** OWASP ZAP provides additional tools that can help users test and analyze web applications, such as intercepting requests, modifying page content, and more. OWASP ZAP is free and open licensed and can be used on any platform – Windows, MacOS, and Linux. It is also highly customizable and extensible, allowing users to customize and adapt it to their needs.

Additionally, OWASP ZAP has reasonable speed and an easy-to-use interface, making it accessible to widespread use. Thanks to an active community of security researchers, OWASP ZAP is constantly updated, and add-ons provide enhanced capabilities for users. OWASP ZAP also integrates with other security testing tools such as Burp Suite.

9.7 Summary

Communication network testing techniques are critical to establishing and maintaining a secure and efficient communications network. Modern technologies, including IoT and cloud computing, continue to create new testing challenges, requiring a more comprehensive and advanced approach to verifying the security and performance of communication networks. The chapter summarizes the possible tests and tools that can be used to evaluate communication networks for telepresence applications.

References

[1] “Iperf,” *Wikipedia*. [Online]. Available: https://en.wikipedia.org/wiki/Iperf. [Accessed: 20-Apr-2024].

[2] “JPerf: A Guide to Measuring Bandwidth Between Two Computers.” [Online]. Available: https://www.informatique-mania.com/ru/linter

net/jperf-manuel-pour-mesurer-la-bande-passante-entre-deux-ordinateu rsslug/. [Accessed: 20-Apr-2024].

[3] "Network Visibility Anywhere." [Online]. Available: https://www.pingpl otter.com/. [Accessed: 20-Apr-2024].

[4] "NetCrunch Monitor, Alert, Visualize," *adremsoft*. [Online]. Available: https://www.adremsoft.com/. [Accessed: 20-Apr-2024].

[5] "Nmap." [Online]. Available: https://nmap.org/. [Accessed: 20-Apr-2024].

[6] LukaSafonov, "Burp Suite: The Swiss Army Knife of Web Application Testing," 2017. [Online]. Available: https://habr.com/ru/articles/328382/. [Accessed: 04-Winter-2024].

[7] "OWASP ZAP overview. Scanner for searching for vulnerabilities in web applications," 2023. [Online]. Available: https://habr.com/ru/companies/first/articles/709586/. [Accessed: 20-Apr-2024].

10

Research Directions

10.1 Introduction

Telepresence, holographic communications, and the metaverse represent a new age in human connectivity and digital interaction. As we approach the technological frontier, the promise of innovation and advancement looms large, providing an unparalleled opportunity to reimagine how we interact, cooperate, and immerse ourselves in virtual environments. The combination of telepresence, holographic communications, and the metaverse marks a paradigm shift in how we perceive and interact with faraway settings and people. It is a convergence driven by rapid advances in display technology, artificial intelligence, and network infrastructure, opening the way for immersive experiences that cross the line between physical and digital realities.

In this age of interconnection, telepresence arises to overcome geographical limits and foster real-time communication and cooperation across great distances. Telepresence allows people to inhabit virtual avatars and connect with others regardless of physical location. Furthermore, holographic communications promise to take telepresence to new heights by displaying lifelike holographic representations of faraway participants, encouraging a sense of presence and immersion previously unavailable through regular video conferencing. This combination of telepresence and holographic technology improves communication and offers new possibilities in various domains, including telemedicine, education, and entertainment.

Furthermore, the metaverse's emergence as a shared, enduring virtual environment opens endless possibilities for exploration, innovation, and social interaction. The metaverse goes beyond the limitations of individual virtual experiences, providing a unified and interconnected digital cosmos in which people can dwell, interact, and create in real time. From immersive virtual worlds and social VR platforms to decentralized virtual economies and user-generated content ecosystems, the metaverse is a vision of interconnected digital domains that reflect the complexity and diversity of the actual world. As we embark on this trip into the metaverse, identity, ownership, and governance concerns arise, requiring us to manage the ethical, legal, and sociological ramifications of a digital existence.

Many research directions await to fully realize the potential of telepresence, holographic communications, and the metaverse. The possibilities are broad and varied, from improving the realism and quality of holographic projections to investigating multi-sensory telepresence experiences that engage all senses. Furthermore, as we explore deeper interactive holographic environments and collaborative virtual spaces, the importance of interdisciplinary collaboration and cross-cutting innovation becomes more apparent. We can unlock the transformative power of these technologies through collaborative research and visionary exploration, resulting in a future where distance is no longer a barrier and the boundaries between physical and virtual realities dissolve into a seamless tapestry of digital experiences. In the rest of this chapter, we discuss the potential research directions of telepresence, holographic communications, and metaverse.

10.2 Holographic Telepresence

Holographic telepresence aims to bridge physical distances between people by projecting realistic holographic representations of remote participants. Improvements in this area could include increasing the fidelity and realism of holographic displays. Holographic displays frequently have limited viewing angles and resolution, resulting in unsatisfactory experiences. Researchers could work on improving holographic display technologies with wider viewing angles, higher resolutions, and better color fidelity. For example, advances in volumetric displays, which generate 3D pictures by illuminating a volume of space, hold promise for improving the realism of holographic telepresence [1].

This includes reducing latency in holographic telepresence systems, resulting in more lifelike and responsive interactions [2]. High latency can interfere with the sensation of presence and immersion, making it difficult for users to engage in fluid conversations or collaborative tasks. Techniques like predictive rendering and network optimization could be researched to reduce latency and improve the overall user experience. Furthermore, incorporating depth sensing and 3D scanning technologies into holographic telepresence systems could allow for real-time reconstruction of faraway settings and people, increasing the impression of spatial presence.

10.3 Interactive Holographic Environments

Interactive holographic environments are designed to allow users to engage with virtual objects and situations through natural gestures and movements. Research in this field could concentrate on building sophisticated gesture recognition algorithms that can accurately read various hand gestures and body motions in real time. Deep neural networks and other machine learning approaches could be used to train gesture recognition models on enormous datasets of annotated motion data. Furthermore, investigating the integration of haptic feedback technology into interactive holographic worlds may improve the sense of touch and manipulation. Haptic feedback devices (e.g., gloves or exoskeletons) may give users tactile sensations when interacting with virtual objects [3, 4].

Researchers could investigate approaches for mimicking realistic physics and dynamics in holographic environments to produce more immersive interactions. Physics-based simulation engines could simulate the behavior of virtual objects and environments, allowing users to interact with them naturally and intuitively. When users interact with virtual objects, they may see realistic collisions, deformations, and gravitational effects. Furthermore, investigating the integration of eye-tracking technologies into interactive holographic worlds may result in more intuitive and responsive interactions. By properly tracking users' sight, holographic systems may anticipate their intentions and alter the presentation of virtual content accordingly.

10.4 Multi-sensory Telepresence

Multi-sensory telepresence seeks to expand visual and aural communication by introducing additional sense modalities such as touch, smell, and taste into remote encounters. Investigating ways to introduce other sensory modalities, including touch, smell, and even taste, into telepresence systems to create a more immersive environment. This could include looking into technologies like haptic feedback devices, smell emitters, and gustatory interfaces. This research could create haptic feedback devices that can simulate genuine tactile experiences. Haptic feedback gloves, for example, might simulate the experience of touching and controlling virtual objects by imparting force and vibration to the users' hands. Advanced haptic feedback algorithms could imitate a variety of textures, surfaces, and physical interactions, increasing the sense of presence and immersion in remote locations [5].

Furthermore, researchers could investigate approaches for simulating olfactory and gustatory sensations in telepresence systems. Olfactory displays, which emit fragrances or odors corresponding to virtual surroundings, may improve the sensation of immersion by activating viewers' sense of smell. Similarly, gustatory interfaces, which stimulate the taste buds to imitate various flavors, may allow users to enjoy culinary delights or beverages from a distance. Creating compact and adjustable olfactory and gustatory display devices may be a substantial research challenge, necessitating advances in smell delivery mechanisms and taste stimulation approaches.

10.5 Holographic Collaboration Tools

Holographic collaboration solutions attempt to improve distant collaboration and communication by providing shared virtual workspaces and interactive content production. Research in this field could concentrate on creating intuitive and user-friendly interfaces for navigating and interacting with holographic environments. User studies and human–computer interaction research could be used to uncover design concepts and interaction paradigms that improve usability and productivity in collaborative environments. For example, investigating natural language processing and speech recognition technologies could allow users to engage with holographic systems through spoken commands, avoiding sophisticated manual input [6].

Moreover, researchers could investigate approaches for enabling real-time collaboration and co-authoring digital content in holographic settings. Collaborative editing tools like shared whiteboards and 3D modeling software could allow users to collaborate effortlessly from multiple places. Developing synchronization algorithms and dispute resolution mechanisms for multi-user interactions in holographic environments may be significant scientific problems. Additionally, investigating the integration of VR and AR technology into holographic collaboration tools could open new avenues for improving distant teamwork and creativity. Mixed-reality experiences combining holographic displays with immersive VR or AR settings may allow users to examine and edit large datasets or architectural ideas interactively [7].

10.6 Metaverse Infrastructure

Researching the infrastructure and networking requirements for building a scalable and interoperable metaverse. This could involve developing protocols and standards for communication, data exchange, and interoperability between virtual worlds and environments. The creation of a reliable and scalable infrastructure is critical to implementing the metaverse. This includes investigating various areas of network architecture, data storage, and communication protocols designed to support immersive virtual environments and integrated digital experiences. Scalability, interoperability, and low-latency communication are critical difficulties that must be addressed to allow seamless interaction and collaboration across the metaverse [8]. Research may concentrate on distributed computing paradigms such as edge computing and peer-to-peer networking to efficiently handle the computational demands of virtual worlds while minimizing latency and assuring high availability.

10.7 Metaverse Governance

As the metaverse expands in complexity and significance, problems about governance, regulation, and digital rights management become more important. This research could examine new governance structures and regulatory frameworks customized to the metaverse's difficulties. This involves intellectual property rights, digital asset ownership, privacy protection, and content moderation. Decentralized governance methods, blockchain-based identity solutions, and cryptographic protocols may provide new pathways for assuring transparency, accountability, and user empowerment inside the metaverse. Furthermore, interdisciplinary study in law, economics, and computer science is required to traverse metaverse governance's legal and ethical complexity.

10.8 Avatar Customization and Representation

Avatar customization is critical to molding users' identities and experiences in the metaverse. The research in this direction includes enhanced avatar creation tools, procedural generation approaches, and machine learning algorithms for creating individualized avatars that accurately reflect users' physical attributes, preferences, and emotions. This includes creating accessible and user-friendly avatars, customizing interfaces, and utilizing data-driven ways to produce diverse and inclusive representations automatically [9]. Furthermore, research may concentrate on improving the realism and expressiveness of avatars using techniques such as facial animation, body language synthesis, and emotion identification. Avatar modification promotes a sense of ownership and agency in the metaverse by allowing users to construct and personalize their virtual identities.

10.9 Social Dynamics and Behavior in Virtual Worlds

Understanding virtual world users' social dynamics and behavior is critical for building lively and inclusive online communities [10]. This research may use social psychology, sociology, anthropology, and human–computer interaction to investigate how people interact, communicate, and collaborate in virtual settings. This includes investigating social norms, group dynamics, identity creation, and social influence in digital environments. Furthermore, this research includes the effects of design decisions, interface affordances, and community management tactics on user engagement, trust, and well-being. Researchers can use insights into the fundamental mechanisms that drive social interaction in virtual worlds to help develop more engaging, helpful, and inclusive online settings.

10.10 Integration with Emerging Technologies

Integrating telepresence, holographic communications, and the metaverse with developing technology opens new opportunities for creativity and synergistic potential. Research in this area includes combining AR and VR technologies to improve the immersive aspects of telepresence and holographic experiences. For example, combining holographic displays and AR glasses could allow users to effortlessly integrate virtual and physical aspects in their surroundings, resulting in mixed-reality experiences that enhance their perception of reality. Furthermore, using AI and ML algorithms can improve the intelligence and interactivity of virtual avatars and settings. AI-powered virtual assistants could help users navigate virtual worlds, complete tasks, and provide context-relevant information [11].

Researchers may look into the potential combinations between telepresence, holographic communications, and other developing technologies like blockchain and the IoT. Integrating blockchain technology into the metaverse could allow for safe digital transactions, decentralized asset ownership, and verifiable identity management. Smart contracts running on blockchain technologies could help automate transactions and enforce digital rights in virtual environments. Furthermore, investigating the integration of telepresence and holographic communications with IoT devices can result in immersive and interactive experiences in smart surroundings. For example, holographic interfaces could be used as intuitive control interfaces for connected devices, allowing people to interact with IoT sensors and actuators in real time.

Furthermore, research may focus on using telepresence and holographic technology to improve distant collaboration in areas such as telemedicine, education, and industry. By merging high-fidelity holographic displays with modern teleoperation technologies, medical professionals can collaborate on difficult surgical procedures, share experiences, and train healthcare staff in real time. Similarly, holographic telepresence technologies can transform distance learning by allowing interactive lectures, virtual field trips, and hands-on laboratory activities. Furthermore, remote specialists can use holographic communication technologies in industrial settings to guide and advise on-site workers, thereby minimizing downtime and increasing production. Researchers who embrace the combination of telepresence, holographic communications, and upcoming technologies can open new avenues for improving human communication, collaboration, and creativity in the digital age.

10.11 Proposed Research of MEGANET 6G Lab

In MEGANET 6G Lab, we have been working in telepresence and HTC setup and testing for three years. In this section, we share a part of our innovative work related to the previously introduced research directions of telepresence and holographic applications.

A. **HolNetVerse**
Based on research in the MEGANET LAB 6G laboratory at St. Petersburg State University of Technology, prof. M. A. Bonch-Bruevich managed to create terminal devices based on octahedra and cylinders, in which the required transmission speed was reduced by more than 1000 times – to 1 Gbit/s. This made it possible not only to come close to the mass implementation of holographic type interactions on communication networks but also to begin work on creating a holographic network universe HolNetVerse (Holographic Network Verse).

B. **Prototype of a moving robot-avatar model for providing telepresence services**
The mobile avatar robot prototype represents an important step forward in telepresence technology. By seamlessly combining a HoloLink holographic teleconference system with a mobile robot, it provides a versatile method of remote communication. This combination of cutting-edge technology enables the transmission of not only holographic pictures but also high-quality audio and kinesthetic data over long distances. Such capabilities have the potential to transform a variety of industries, including healthcare, education, business, and entertainment.

In terms of expecting the future of communication networks, the development of this prototype is consistent with the trajectory set forth by IMT-2030 network services. With a focus on the HolNetVerse idea, it aims to bridge the divide between urban and rural areas, ensuring fair access to advanced communication technology. By delivering telepresence capabilities to underserved areas, the project hopes to empower communities by encouraging virtual connections that cross geographical barriers. This not only improves connectivity but also creates chances for collaboration, education, and socioeconomic development.

The implications of deploying such mobile avatar robots extend far beyond conventional telecommunication. In healthcare, for instance, they could enable remote medical consultations with specialists, allowing patients in remote areas to access expertise without the need for extensive travel. Similarly, in education, students could benefit from virtual field trips guided by experts or interactive lessons facilitated by teachers from around the world. Furthermore, in industries such as manufacturing and maintenance, these robots could provide remote assistance, reducing downtime and increasing efficiency. Moreover, the integration of the HoloLink holographic teleconference system opens avenues for immersive experiences previously unimaginable. Whether it is attending a conference as a holographic presence or participating in virtual trade shows and exhibitions, individuals can engage in events without the constraints of physical presence. This not only reduces carbon footprints associated with travel but also enhances inclusivity by accommodating individuals with mobility limitations or those residing in remote areas.

Thinking about ethical, security, and privacy issues is becoming more important as the mobile avatar robot is being developed. It is critical to guarantee secure communication channels and to safeguard sensitive data as it is transmitted across these systems. Furthermore, telepresence technology needs ethical standards to control its usage, which will handle issues like autonomy, consent, and the fuzziness of the line between the actual and virtual worlds. To benefit from telepresence's revolutionary potential while reducing its risks, the suggested robot avatar promotes a responsible approach to innovation. Figures 10.1 and 10.2 present two of our proposed robot avatar systems.

Figure 10.1 MEGANET 6G Lab's proposed robot avatar in action with humans.

Figure 10.2 MEGANET 6G Lab's proposed robot avatar with hologram generation.

C. Prototype of a tactile suit for providing telepresence services

Figures 10.3– 10.5 present the proposed tactile suit under testing for telepresence services. The suit represents a paradigm shift in how we interact with virtual and augmented reality environments. It goes beyond typical input devices by providing an integrated hardware system that works smoothly with these immersive situations. The suit represents the convergence of numerous technologies, including the Internet of Skills, IoT, and the Tactile Internet. The suit's functionality is centered on its numerous sensors, which have been precisely engineered to capture and interpret a vast amount of data. These sensors create a full representation of the user's physical movements and interactions, transforming them into meaningful actions in virtual and augmented environments. Unlike traditional input devices, which frequently rely primarily on hand-held controllers or gesture recognition, Suit takes a comprehensive approach to immersion, incorporating every movement, touch, and gesture into the user's virtual experience.

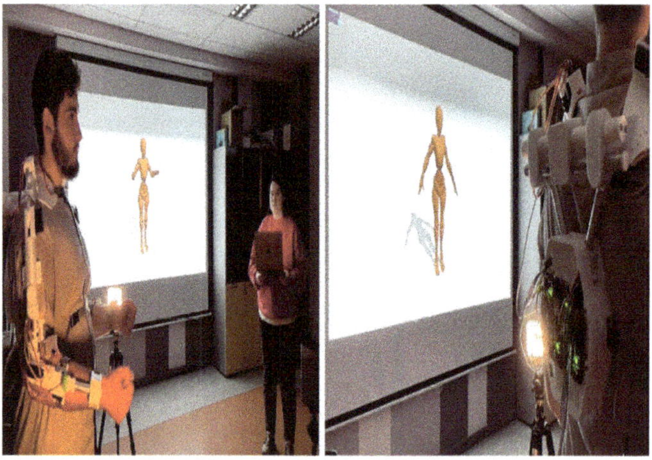

Figure 10.3 MEGANET 6G Lab's proposed tactile suit.

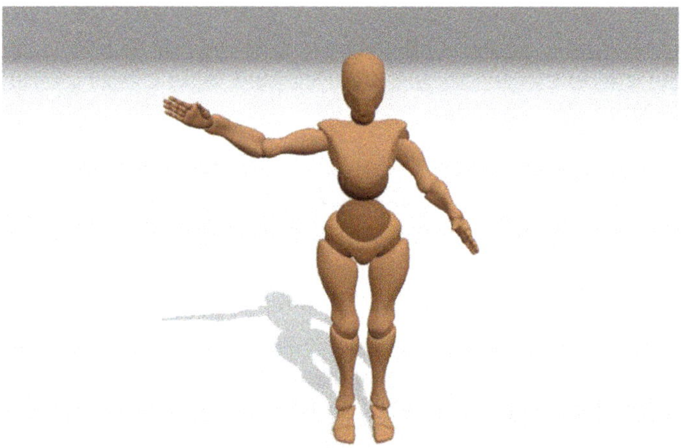

Figure 10.4 3D model of the MEGANET 6G Lab's proposed tactile suit in action.

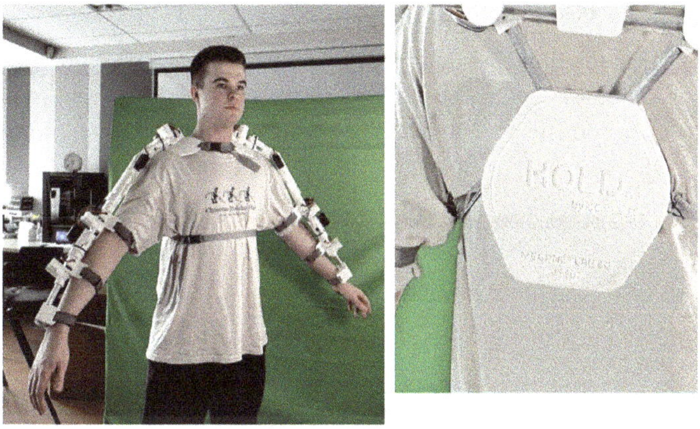

Figure 10.5 Front and back views of MEGANET 6G Lab's proposed tactile suit in action.

Furthermore, Suit is more than simply an interface; it is a means of embodying digital avatars in virtual spaces. The suit collects kinesthetic data via its complex network of sensors, thereby duplicating the user's physical activities in the digital realm. This seamless integration of the physical and virtual realities increases users' sense of presence and agency, allowing them to traverse and interact with virtual worlds as if they were extensions of their physical selves.

In terms of telepresence, the suit has enormous promise for bridging the gap between remote individuals and organizations. Users can connect remotely with unprecedented realism by inhabiting digital avatars equipped with Suit technology. The suit's capacity to send kinesthetic data in real time enables sophisticated conversation, with gestures, body language, and tactile feedback enhancing the telepresence experience. Suit redefines how we interact and communicate across distances, whether it is through project collaboration, virtual meetings, or exploration of shared virtual places.

A wide range of industries, including media, education, healthcare, and more, stand to benefit from Suit's ongoing development. The boundaries between fact and fiction are becoming increasingly porous in the entertainment industry because of increased opportunities for interactive experiences and immersive storytelling. One way it can improve education is by allowing students to have more active roles in their learning using virtual environments. With its focus on physical movements and interactions, Suit has great potential for remote rehabilitation and therapy in the healthcare industry. All things considered, Suit marks the beginning of a new age in human–computer interaction, one in which the lines between the real and virtual worlds blur, opening previously unthinkable avenues for sharing ideas, working together, and being creative.

References

[1] V. Poulkov *et al.*, "The HOLOTWIN project: Holographic telepresence combining 3D imaging, haptics and AI," in *2023 Joint International Conference on Digital Arts, Media and Technology with ECTI Northern Section Conference on Electrical, Electronics*, IEEE, 2023, pp. 537–541.

[2] R. Cheng, K. Liu, N. Wu, and B. Han, "Enriching telepresence with semantic-driven holographic communication," in *Proceedings of the 22nd ACM Workshop on Hot Topics in Networks*, 2023.

[3] F. E. Fadzli, A. W. Ismail, G. Su, and S. Sadimon, "Evaluation of User Interaction Technique with Holographic Projection in Extended Reality," in *International Conference on Trends in Sustainable Computing and Machine Intelligence*, Singapore; Singapore: Springer Nature, 2023, pp. 243–258.

[4] A. Adilkhanov, M. Rubagotti, and Z. Kappassov, "Haptic devices: Wearability-based taxonomy and literature review," *IEEE Access*, vol. 10, pp. 91923–91947, 2022.

[5] B. Han and H. D. Schotten, "Multi-sensory HMI for human-centric industrial digital twins: A 6G vision of future industry," in *2022 IEEE Symposium on Computers and Communications (ISCC)*, 2022, pp. 1–7.

[6] C. Bolognesi and D. Sorrenti, "New Representation Tools in VR and Holographic View," in *Beyond Digital Representation: Advanced Experiences in AR and AI for Cultural Heritage and Innovative Design*, Cham; Nature Switzerland: Springer, 2023, pp. 805–820.

[7] Y. Li, Q. Yang, J. Xiong, K. Yin, and S.-T. Wu, "3D displays in augmented and virtual realities with holographic optical elements [Invited]," *Opt. Express*, vol. 29, no. 26, p. 42696, 2021.

[8] S. Cheng, "Basic Infrastructure of the Metaverse," in *Metaverse: Concept, Content and Context*, Cham: Springer Nature Switzerland, 2023, pp. 25–46.

[9] R. Ratan, M. S. Klein, C. R. Ucha, and L. L. Cherchiglia, "Avatar customization orientation and undergraduate-course outcomes: Actual-self avatars are better than ideal-self and future-self avatars," *Comput. Educ.*, vol. 191, no. 104643, p. 104643, 2022.

[10] G. Freeman, S. Zamanifard, D. Maloney, and A. Adkins, "My body, my avatar: How people perceive their avatars in social virtual reality," in *Extended Abstracts of the 2020 CHI Conference on Human Factors in Computing Systems*, 2020.

[11] H. Yu, T. Taleb, K. Samdanis, and J. Song, "Toward supporting holographic services over deterministic 6G integrated terrestrial and non-terrestrial networks," *IEEE Netw.*, vol. 38, no. 1, pp. 262–271, 2024.

Index

About the Authors

Abdelhamied A. Ateya (Senior Member, IEEE) received his B.Sc. and M.Sc. degrees in electrical engineering from Zagazig University, Egypt, in 2010 and 2014, respectively, and his Ph.D. degree from the Saint-Petersburg State University of Telecommunications, Russia, in 2019. He is currently an Assistant Professor at the Department of Electronics and Communications Engineering, Faculty of Engineering, Zagazig University. He has coauthored more than 50 publications in high ranked journals. His current research interests include machine learning applications in communication networks, 5G/6G communications, the Internet of Things, tactile internet and its standardization, and vehicular communications. He is a member of many scientific communities and ACM Prof. Member. He has been an active member of several international journals and conferences, with a contribution as an author, a reviewer, an editor, or a member of program committees.

Ahmed A. Abd El-Latif (Senior Member, IEEE) received his B.S. and M.S. degrees in computer science from Menoufia University, Egypt, and his Ph.D. degree in computer science from the Harbin Institute of Technology, Harbin, China. Currently, he holds staff positions at Menoufia University, Egypt, and Prince Sultan University, Saudi Arabia. He has published over 300 papers in quality journals and conference proceedings, including 15 books, with over 11300 citations. He is Head of the MEGANET 6G Lab Research in the Russian Federation. He is the founder and deputy director of the Center of Excellence in Quantum and Intelligent Computing (Prince Sultan University, Saudi Arabia). He has received many awards, including the State Encouragement Award in Engineering Sciences from Egypt in 2016, the Best Ph.D. Student Award from Harbin Institute of Technology, China, in 2013, and the Young Scientist Award from Menoufia University, Egypt, in 2014. He is actively engaged in the scientific community, serving as Chair or Co-Chair of several conferences and holding editorial positions in various quality journals.

Ammar Muthanna is an Associate Professor at the Department of Telecommunication networks, SPbSUT, and Director of the Scientific Center for Future Wireless Networks Modeling at RUDN University, Institute of Applied Mathematics and Telecommunications, RUDN University. He received his B.Sc., M.Sc., Ph.D., and Dr.Sc. degrees from the Saint-Petersburg State University of Telecommunications in 2009, 2011, 2016, and 2023 respectively. In 2017-2019 he worked as Postdoctoral Researcher at RUDN University. In 2012 and 2013, he took part in the Erasmus student Program with the Faculty of electrical engineering, University of Ljubljana, and in 2014 he was visitor researcher at Tampere University, Finland. Ammar is a senior member of the IEEE and ACM member. He has been an active member of the technical program committee at many international conferences and journals. He is the expert on the Judges Panel and the Challenge Management Board at AI-5G-Challenge, ITU, and organizer of the Russian Host in 2020 and 2021 at Saint-Petersburg State University of Telecommunications. His research interests include wireless communications, 5G/6G cellular systems, artificial intelligence, high-density IoT, telepresence applications, edge/fog computing, and software-defined networking.

Artem Volkov is an Associate Professor of the Saint-Petersburg State University of Telecommunications (SPbSUT) and head of experimental development in Meganetlab 6G. He defended his Ph.D. thesis "Future Infrastructure & Services Based on AI" in 2021 in Russian Federation. He is the author of scientific articles that are indexed in the Scopus & WoS databases and he also co-authored the book "Software-define Networking" in the Russian language. He specializes in 5G/6G infrastructure issues including SDN networking, MEC/fog computing, and telepresence services, taking into account new approaches of AI implementation to networks. He is also an editor of draft recommendations in ITU-T SG 11, SG 13, and SG 20. He has also participated as a Russian delegate in the International Telecommunication Union Standardization Sector.

Andrey Koucheryavy received his degree from the Leningrad University of Telecommunications in 1974. He joined the Telecommunication Research Institute LONIIS, where he worked until October 2003 (from 1986 to 2003 as the First Deputy Director). He has been a Professor with the Bonch-Bruevich Saint Petersburg State University of Telecommunications (SUT) since 1998, where he became a Chaired Professor with the Telecommunication Networks and Data Transmission Department in 2011. He was an Advisor of the Central

Science Research Telecommunication Institute (ZNIIS) from 2003 to 2010. He was a Co-Founder of the International Teletraffic Seminar in 1993, 1995, 1998, and 2002; the Founder of the Model Network for Digital Networks with LONIIS in 1997; the Co-Founder of the Model Network for Packet Networks with ZNIIS in 2004, the Internet of Things Laboratory in 2012, and the Quality of Experience and IPTV Laboratory with SUT in 2014. He was the Chair of the Scientific School on Teletraffic Theory in LONIIS from 1990 to 2003; the Founder and the Scientific School Chair of the Internet of Things and Self-Organizing Networks with SUT since 2010; a Steering Committee Member of IEEE technically co-sponsored series of conferences ICACT and NEW2AN. He was the SG11 ITU-T Vice-Chairperson, from 2005 to 2008 and from 2009 to 2012. He was the WP3/WP4 SG11 Chairperson from 2006 to 2012; the WP4 SG11 Vice-Chairperson, from 2015 to 2016; the Chairperson of SG11 from 2017 to March 2022. He was the Co-Founder of the International Testing Center for New Telecommunications Technologies with ZNIIS under ITU-D Competence. He was a Host and a Technical Program Committees Member of the "Kaleidoscope 2014" at SUT. He was an Honorary Member of Popov's Society in 2002.

For Product Safety Concerns and Information please contact our EU
representative GPSR@taylorandfrancis.com
Taylor & Francis Verlag GmbH, Kaufingerstraße 24, 80331 München, Germany

www.ingramcontent.com/pod-product-compliance
Ingram Content Group UK Ltd.
Pitfield, Milton Keynes, MK11 3LW, UK
UKHW022210190926
475702UK00003B/291

* 9 7 8 8 7 7 0 0 4 6 7 3 2 *